Rehabilitation Approach in Autism

Manu Goyal • Kanu Goyal
Editors

Rehabilitation Approach in Autism

Springer

Editors
Manu Goyal
Maharishi Markandeshwar Institute
of Physiotherapy and Rehabilitation
Maharishi Markandeshwar
(Deemed to be University)
Mullana - Ambala, Haryana, India

Kanu Goyal
Maharishi Markandeshwar Institute
of Physiotherapy and Rehabilitation
Maharishi Markandeshwar
(Deemed to be University)
Mullana - Ambala, Haryana, India

ISBN 978-981-96-4161-1 ISBN 978-981-96-4162-8 (eBook)
https://doi.org/10.1007/978-981-96-4162-8

This Springer imprint is published by the registered company Springer Nature Singapore Pte Ltd.
The registered company address is: 152 Beach Road, #21-01/04 Gateway East, Singapore 189721, Singapore

Foreword

Autism spectrum disorder (ASD) presents a diverse array of challenges that vary from child to child. With no two children experiencing autism the same way, a one-size-fits-all approach is insufficient for effective management. As our understanding of autism evolves, so does the recognition that a comprehensive, multi-disciplinary approach is essential. This book brings together two critical elements: new approaches to managing autism and the importance of a team-based approach, offering professionals practical strategies for holistic care.

Autism affects multiple areas of development, including communication, behavior, sensory processing, and social interactions. Given its complexity, no single professional can address all aspects of a child's needs. The collaboration of these professionals ensures that care is tailored to the unique strengths and challenges of each child. Role of team-based autism care and management has been elaborated in one of the chapters in this book.

Role of occupational therapy is and always shall be of paramount importance, as highlighted in a chapter in book. The other chapters highlight some of new approaches, e.g., hands on approach, aquatic therapy, play therapy, etc. in autism management. While these new approaches are promising, their success depends on being delivered within a coordinated, multi-disciplinary framework. Also, as the child grows, their needs change, and the team must adapt its strategies to ensure ongoing support through the school years and into adulthood. An essential aspect of the team approach is the inclusion of the "family." Parents and caregivers, with their intimate knowledge of their child's behavior, play a crucial role in shaping the care plan. Their input ensures that the team's interventions are aligned with the child's daily experiences and long-term goals. This partnership between professionals and families fosters a supportive environment in which the child's strengths can be nurtured.

In summary, this book *A Rehabilitation Approach in Autism* underscores the transformative power of combining innovative approaches with a collaborative team effort. By working together and integrating new techniques with proven strategies, professionals can better meet the complex needs of children with autism, helping them thrive. I wish the readers of this book a happy and informative experience of learning autism management.

Professor & Head, Pediatrics
Maharishi Markandeshwar Institute
of Medical Sciences & Research
Maharishi Markandeshwar (Deemed to be University)
Mullana - Ambala, Haryana, India

Baljeet Maini

Preface

Rehabilitation in Autism-Spectrum-Disorder (ASD) is rapidly evolving, driven by ongoing research and in-depth understanding of the complexities of the condition. This book aims to serve a comprehensive source for researchers, practitioners, and families affected by autism, providing insights into effective rehabilitation assessment, strategies, and interventions. This book will explore the multifactorial aspects of ASD and their implications for rehabilitation. The book chapters in *A Rehabilitation Approach in Autism* will address various dimensions of ASD rehabilitation, including evidence-based-practices, individualized approaches, and interdisciplinary collaborations. As we delve into the intricacies of the rehabilitation in autism, it is imperative to remain aware about the latest research findings regarding potential therapeutic targets and various innovative rehabilitation aspects. This book aims to inform and inspire the readers to engage with the latest developments in autism research and practice. I express my heartfelt gratitude, to all the contributors for their invaluable insights and dedication toward the completion of the book. It's a call to action for more compassionate, comprehensive care for those affected. Welcome to a resource that we hope will enlighten, encourage, and empower you or your loved ones to navigate the complexities of ASD with confidence and hope.

Mullana - Ambala, Haryana, India

Manu Goyal
Kanu Goyal

Contents

An Overview of Autism

1

Parul Sharma, Shweta Sharma, and Urvi

1.1 Introduction

Autism spectrum disorder (ASD) is a neurodevelopmental condition marked by limited interests, repetitive activities, and difficulties in social interactions [1]. Early diagnosis is crucial for improving behavioral problems, functional skills, and communication difficulties in autism. The International Classification of Diseases (ICD-10) and Diagnostic and Statistical Manual of Mental Disorders (DSM IV) classify autism under pervasive developmental disorder (PDD) [2] and the Diagnostic and Statistical Manual of Mental Disorders, 4th edition (DSM IV) as Autism [3]. Recently, DSM-5 (American Psychiatric Association [4]) and ICD-11 (World Health Organization [5]) have classified autism into autism spectrum disorder (ASD). It is widely recognized that Rett syndrome is a distinct neurological disorder. Furthermore, individuals who struggle with interpersonal interaction but do not exhibit persistent or confined behavioral patterns may be diagnosed with social pragmatic communication disorder (SPCD), a recently developed illness. Different severity categories of ASD are used to gauge how much help a person needs [6].

P. Sharma (✉)
School of Physiotherapy, Delhi Pharmaceutical Sciences and Research University, New Delhi, India

S. Sharma
Maharishi Markandeshwar Institute of Physiotherapy and Rehabilitation, Maharishi Markandeshwar (Deemed to be University), Mullana - Ambala, Haryana, India
e-mail: shweta.sharma@mmumullana.org

Urvi
Faculty of Physiotherapy, SGT University, Gurugram, Haryana, India

M. Goyal, K. Goyal (eds.), *Rehabilitation Approach in Autism*,
https://doi.org/10.1007/978-981-96-4162-8_1

1.2 Epidemiology

Recent studies show a significant increase in the prevalence of ASD, with boys experiencing a four to five times higher rate. The average prevalence in Asia, Europe, and North America is 1%. In the United States, the prevalence was 2.5% in 2014–2016, while in Asia, it was 3.9% [7]. The increase in ASD is largely due to increased awareness and awareness, but it is largely real [8]. The average age of diagnosis for ASD is 4–7 years, with delays more pronounced in lower-income, ethnic/racial minority, and rural children. White children and those with higher socioeconomic status are more likely to be diagnosed earlier than Black, Latinx, Asian, and low-income children. This results in non-White and lower-income children not benefiting from early intervention services during crucial developmental windows [9].

A recent meta-analysis found that the male-to-female ratio for ASD is closer to 3:1 than previously thought [10–12] (Fig. 1.1) and females who meet the criteria are more likely to go undiagnosed. This could be due to the autism phenotype in women contributing to misdiagnosis, delayed diagnosis, or missed diagnosis in girls [13]. Additionally, gender biases and prejudices may hinder early diagnosis of ASD in girls [14].

1.3 Factors

ASD is a neurobiological disorder influenced by genetic and environmental factors. Research is ongoing to identify potential etiologic pathways; nonetheless, a single reason has not been found.

Neuropathologic investigations have identified malformations, aberrations in the limbic system, variations in cerebellar architecture, and changes in frontal and temporal lobe cortical modifications [15–17]. A study of young children's neocortical architecture revealed focal alteration of the cortical laminar architecture, suggesting difficulties with the growth of cortical layers and the development of neurons [18].

Children with ASD exhibit brain overgrowth in cortical size and extra-axial fluid, a study aiming to understand the disorder's etiology and potential biomarker [19, 20]. Genetic factors influence ASD susceptibility, with siblings having higher diagnosis risk and monozygotic twins having more likely diagnoses [21].

Environmental factors, including prenatal, perinatal, and postnatal factors, can influence genetic risk in certain patients [22] (Fig. 1.2). Prenatal folic acid supplements may lower risk in patients exposed to antiepileptic medicines, but exposure to

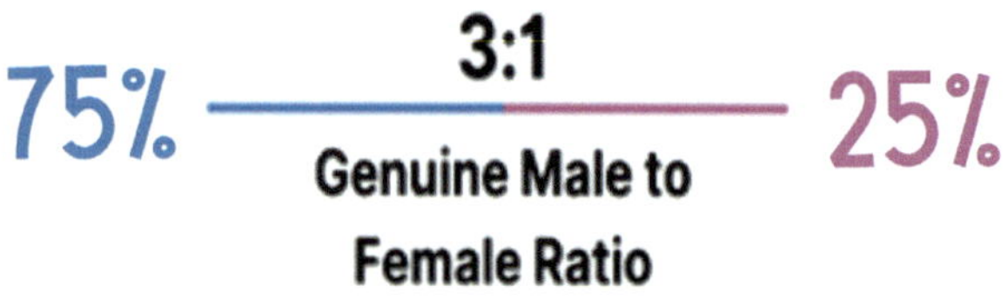

Fig. 1.1 Gender-wise prevalence of ASD. (Source: self)

Fig. 1.2 Factors influencing ASD. (Source: self)

thalidomide and valproic acid raises risk [23]. Research on folic acid in autism is limited [24], and older parents are more likely to have children with the condition [25]. There is also speculation about autoimmune diseases in the mother, but conflicting findings remain [26]. Recent studies suggest that maternal infection or immunological activation during pregnancy may increase the risk of ASD, with longer and shorter interpregnancy periods increasing the risk [27, 28], and premature babies more likely to experience ASD [29].

1.4 Characteristics Associated with Autism Spectrum Disorder

- *Communication*

Although there are wide variations in language proficiency among individuals, all persons with ASD struggle with language and communication. While some people speak a lot but struggle with social language use, others are nonverbal. It can seem as though those who suffer from autism spectrum disorders live an isolated environment in which communication is not encouraged.

This is more of a communication breakdown than a deliberate action [30].

Among the language challenges are:

- Difficulties with nonverbal communication
- Challenges relating to nonverbal communication:
  - Strange body postures
  - Inappropriate facial expressions

  - Unusual gestures
  - No eye contact
  - Lack of a distinct or shared attentional focus

A delay in or deficiency in expressive language abilities

- Notable variations in oral language for language-developing individuals:
  - Unusual rhythm or stress
  - Unusual pitch or intonation
  - Abnormally fast or slow speech rate
  - Monotonous voice quality
- A propensity to use language for social rather than for meeting needs
- Peculiar and repetitive speech patterns [31]
- Echolalic speech: Repeating someone else's words, whether immediately or over time, may indicate an attempt to communicate, indicate speech production and imitation, and potentially aid a pupil with their communicative or cognitive needs.
- Restricted vocabulary: The text is primarily composed of nouns, often used to describe physical responses, and has constraints in the social responsibilities.
- The tendency to persist on a specific topic, characterized by persistent discussion and difficulty in changing it.
- Starting a discussion, adhering to unspoken guidelines, maintaining it progressing, inappropriately interrupting, being rigid in one's approach, and speaking in stereotypes are all examples of communication problems.

Students with ASD struggle with verbal information comprehension, following long instructions, and remembering sequences, with difficulty varying among individuals and even high-functioning individuals with normal intelligence [31] (Fig. 1.3).

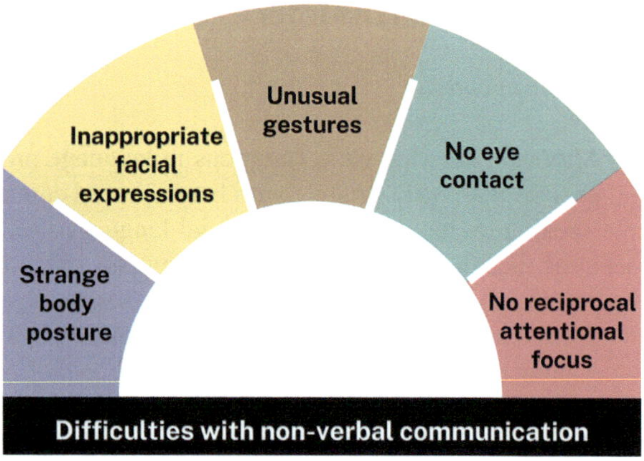

Fig. 1.3 Various difficulties in nonverbal communication. (Source: self)

Implications

Entire communication evaluation and treatment are part of strategies for students with pervasive developmental disorders and ASD, including speech-language pathology, informal observation, and classroom evaluation, to identify goals and strategies for language skill development.

Instruction should emphasize: This include being focused, replicating, understanding commands and common phrases, utilizing language in social contexts, and engaging in meaningful conversation. Communication strategies should focus on how effectively language and communication can be used in different situations.

- *Social Interaction*

Students with ASD often have difficulty forming and keeping relationships, and they exhibit qualitative differences in social interaction. They might only engage in a restricted social circle or adopt a rigid social interaction style. Challenges arise from inability to extract social information from social interactions and respond effectively, not out of a lack of enthusiasm or a hesitation to interact with others [32].

Individuals with ASD often struggle with language processing and nonverbal communication, which are crucial for understanding social situations. They may miss important cues like tone of voice and facial expressions, struggle to read nonverbal cues, and use nonverbal behaviors [33].

Pupils diagnosed with ASD frequently struggle to comprehend other people's viewpoints, let alone recognize that those viewpoints might differ from their own [34]. They might also struggle to comprehend the views, intentions, desires, knowledge, and perceptions of both themselves and other people. They might struggle to comprehend how thoughts and deeds are related. For instance, ASD children may struggle to understand the sadness of others, even if they are not experiencing the same sadness themselves.

- *Learning*

Uneven patterns of development are a defining feature of the psychoeducational profiles of students with ASD. Research suggests that while many cognitive functions may be compromised, not all are. Furthermore, complex abilities may be impaired while simpler abilities in the same domain remain intact. Some of the cognitive traits that are frequently connected to ASD have been identified by current research, and they include the following: Impairments in concept formation, abstract reasoning, receptive and expressive language, social cognition, attention and emotion sharing, and problem-solving can result in difficulties in expressing abstract concepts, sharing attention, understanding others' feelings, and organizing [35].

In comparison to other subjects, some students perform better on rote memory and visual-spatial tasks. In fact, they might be particularly good at matching, spatial, perceptual, and visual-spatial tasks like assembling puzzles. Some people might have trouble remembering more complicated information, but they might be able to recall simpler information. Certain students may find it difficult to learn concepts that cannot be represented by images, while others may learn information more quickly and easily when it is presented in a visual format [36].

ASD patients may face difficulties in understanding both written and oral information, including following instructions and comprehending what they read. However, a small percentage of higher-functioning people possess a reasonable level of word recognition, phonetic proficiency, and word meaning comprehension. Some patients show strength in specific areas of communication, like vocabulary, sound production, and basic formation of grammar, but they struggle greatly when it comes to conversing and using speech in social and interactive contexts. In a similar vein, a high-functioning student might find it difficult to solve mathematical problems but be able to execute numerical computations fairly well.

Implications

A learner's behavior, social interactions, and academic performance follow patterns of strengths and weaknesses as a result of these cognitive differences. Cognitive skills development is uneven, and educational programs should focus on individual needs and strengths. Students with ASD often struggle with concepts, attention, language, and memory retention. Programs may need continuous adjustment. Considering these characteristics in addition to firsthand accounts of visually inclined individuals with ASD, it would seem sense to incorporate visual materials into the curriculum.

- *Unusual Attentional Patterns*

Unusual attention patterns are frequently displayed by students diagnosed with ASD. These challenges affect social development, communication, and academic skill acquisition because they usually find it difficult to give focus to relevant cues or information in their surroundings, and students may concentrate on one part of their surroundings to the absence of other significant information. For instance, a pupil might gaze at a ball but not at the target of the throw. Alternatively, a student might pick up on little details like a staple in a paper's corner but miss the content on it. This is known as over-selectivity to stimuli.

The inability to divide attention equally between two objects or people is another characteristic of this disorder. This has been described as a joint or shared attention problem. For instance, a lot of people with ASD do not try to get other people to notice things or activities that they find interesting. Similar to this, children with ASD frequently overlook things or activities that other people find interesting [35–37].

Pupils may struggle to disengage and switch their focus from one stimulus to the next, which could be a factor in their intolerance and reluctance to adapt. They could also have limited attention skills.

Implications

Attention issues can seriously impede students' capacity to acquire appropriate social behavior and language. For instance, students diagnosed with ASD may react to insignificant social cues that grab their attention or focus on specific parts of a conversation without comprehending the meaning being conveyed. They might miss crucial nuances in the message because they are not paying attention to various speech and language cues.

Students should receive information and learning activities that are straightforward, suitable for their understanding and engaging, and emphasize key details. It is possible to incorporate individualized attention-focused strategies for students into lesson plans. Parents can share important knowledge about how they help their kids concentrate on the things they need to learn. Instructional plans should ideally involve assisting students in eventually managing these strategies on their own.

- *Unusual Responses to Sensory Stimuli*

The sensory experiences of people with ASD are typically different from those of others. From hyposensitivity to hypersensitivity, there are several possible reactions to sensory stimulation. A person may occasionally exhibit either an overly or underreactive response to one or more of their senses. People with ASD may find environmental stimuli upsetting or even painful. This could be true for any or all forms of sensory information.

One possible contributing factor to other traits linked to ASD is a problem with sensory processing. It is unclear how much other traits linked to autism spectrum disorders are influenced by sensory issues. Nonetheless, there is enough data to recommend that the kind and volume of sensory stimulation present in the surroundings, as well as the person's response to it, be taken into account [35, 36].

Types of Sensory Integration Dysfunctions in Autism

Sensory integration disorder, as described by [38], is characterized by three basic patterns and is characterized by challenges with adaptively classifying, regulating, and organizing sensations:

1. Dyspraxia: Comprehending, arranging, arranging, and executing atypical motor action sequences can be challenging while managing one's own body or things. Individuals might be wrongly assumed to be hesitant to labor and to engage in interesting jobs.
2. Sensory discrimination disorder: The challenge of comprehending the spatial and temporal information included in sensory perceptions.
3. Dysfunction of sensory modulation: The ability to gradually and effectively change how you respond to sensory information is known as sensory modulation. Sensory modulation dysfunction occurs when someone struggles to appropriately control their reactions to sensory stimuli [37].

Implications

Some of the inappropriate behaviors exhibited by children with ASD can be attributed to these unpleasant or aversive sensory experiences. Students who suffer from severe sensory processing issues, for instance, may completely shut down in order to avoid unpleasant stimuli or excessive stimulation. The urge to get away from overstimulating situations could be the cause of tantrums.

Self-stimulating behaviors, which produce a self-controlled, repetitive stimulus, may assist people in calming down when external stimuli become too much to handle.

Understanding the various sensory integration and stimulation experiences is crucial to comprehending the behaviors of students with ASD and developing educational plans for them [39].

- *Anxiety*

The DSM-IV does not explicitly mention anxiety in its criteria for ASD, but many individuals with the condition, including parents and teachers, identify anxiety as a trait. Numerous factors can contribute to anxiety, such as:

- A strong need for consistency and trouble with change and transitions
- The inability to communicate oneself
- Difficulties interpreting sensory data
- Sensitivity to stimulation from sensory sources
- Inability to understand social expectations
- Fear of situations that are not understood

Implications

Programming for kids with ASD often addresses anxiety and the aspects that tend to contribute to it. A range of techniques can be employed to assist students in managing their anxiety and coping with challenging circumstances, and adjustments and modifications can be made to the surroundings to lessen anxiety-inducing circumstances. One of these strategies is to use social scripts to promote appropriate soothing techniques or ability to cope.

Other strategies include:

- Alerting people about upcoming changes and transitions
- Creating weekly and daily routines to improve predictability
- Giving accurate information on anxiety- or fear-inducing circumstances, such what to do in a lost case
- Making the classroom a calm environment

Therefore, it is obvious that the learning characteristics associated with ASD vary and that these differences have significant implications for instruction in all subject areas and developmental domains. These variations generally call for a heavy emphasis on methodical, explicit, or “direct” instruction. To put it simply, children who have ASD need a great deal of structure in the way that materials are presented, the learning environment is set up, and the teaching strategies used. With children who have ASD, direct instruction should be the main tool utilized, but it is also critical for teachers to be adaptable and adjust expectations as needed. A successful program must strike a balance between the requirement for meticulously prepared instruction and the ability to modify strategies in response to the students' evolving needs.

1.5 Evaluation

The evaluation process for ASD includes screening all children for risk or symptoms. The American Academy of Pediatrics (AAP) recommends developmental checks at 9, 15, and 30 months, routine well-child visits, and autism-specific screenings at 18, 24, or 30 months. Early signs include low eye contact and lack of reaction, and language and social skills loss [40] are two screening tests for ASD in this demographic. Preschoolers may display unusual behaviors like pretend play, focused hobbies, and stiffness, which may indicate social issues. However, they may also have concrete thinking, struggle with emotions, and show interest in peers. Screening instruments like the SCQ (Social Communication Questionnaire), SRS (Social Responsiveness Scale), and ASSQ (Autism Spectrum Screening Questionnaire) can help identify potential ASD in these groups [41].

Primary care physicians should refer children under 3 years old to early intervention services and those older to the public-school system for a psychoeducational evaluation. For a definitive diagnosis, a referral to a specialist is also needed [42]. A comprehensive evaluation of a child with ASD should include a physical examination for dysmorphic traits, a neurologic examination, a parent interview, observations from outside sources, and clinical observation by a clinician with experience treating ASD [43].

ASD patients typically do not require further laboratory testing beyond genetic testing. However, those with specific findings or risk factors may benefit from a metabolic workup. This may include a metabolic evaluation if the patient exhibits signs of developmental regression, coarse features, organ involvement concerns, seizures, or ataxia. The evaluation may include CBC, liver and renal function tests, and urine organic acids or glycosaminoglycans [44]. Children with pica history should have lead levels measured and nutritional status assessed [45]. Restless sleep symptoms may require sleep study referrals, and iron deficiency evaluations may be necessary, especially if dietary rigidity limits iron intake [46].

Neuroimaging may be suitable for ASD patients with microcephaly, suspicious TSC, or abnormal neurologic exam, while electroencephalography (EEG) should be performed on those displaying seizures [47]. Children with signs of genetic, metabolic, or neurological disorders should be immediately referred to a professional for testing. Routine testing for intestinal permeability studies is recommended. Immunologic or neurochemical markers, mitochondrial disorders, allergies, hair analysis, erythrocyte glutathione peroxidase studies, stool analysis, and vitamin and mineral deficiencies are not recommended.

1.6 Evidence-Based Natural Therapies and Lifestyle Recommendations

The therapist will employ a variety of strategies based on an assessment of the patient's particular needs in conjunction with our current understanding of the pathophysiology that underlies autism. Although most of this research is in its early stages, the following treatments are supported by evidence.

Diet Free from Gluten and Casein (GFCF) Using a gluten-free casein-free (GFCF) diet for 8–12 months resulted in improvements in key autistic behaviors, according to a controlled, single-blind research conducted in Scandinavia in 2010. Adolescent patients had a greater dropout rate, according to this study. Communication, social interaction, focus, attention span, and hyperactivity all showed improvements. After 24 months, there were improvements in stereotyped and repetitive behaviors as well as social interaction [48].

Discrete Trial Teaching (DTT) As the above chart illustrates, the developmental responsive approach is entirely different from discrete trial teaching (DDT). The DTT has the potential benefit of being an excellent tool for teaching new communication techniques [46]. Initially implemented by Lovaas in 1987, the DTT adheres to the conventional behavioral method commonly utilized in the education of children with ASD and has been shown by several researchers to provide favorable results [49–51]. Using this method, an adult might, for instance, offer a discriminative stimulus that combines nonverbal cues (finding a little toy) with verbal cues ("Let's talk about…"). The concept of direct response theory (DTT) suggests that children learn target behaviors when they receive consistent responses from adults. This theory is based on the idea that children learn when they receive clear instructions and support. If a child does not respond, the practitioner provides the correct answer. The teacher may employ verbal instructions and actions if the student still does not. However, children with ASD who struggle socially are not part of this concept's examination [49].

Transactional Approach It combines the benefits of the developing responsive and developmental responsive approaches, providing step-by-step instruction in an interactive game-based setting. It aims to teach students how to initiate joint attention effectively [52]. The method uses enhanced milieu teaching (EMT) and social communication, emotional regulation, and transactional support (SCERTS) [53]. The success of the intervention relies on how responsive the primary social partners are. Transactional techniques and DTT approaches encourage advanced communicative behaviors in students [53, 54].

Combined Approach The elements of the DTT and transactional techniques are integrated in the final strategy, known as the combined approach. Since the pertinent treatments specifically incorporate the elements of these two approaches, they are referred to as the combined approach together. The DTT and transactional approaches are blended in an attempt to create a new way by utilizing distinct elements of each [54, 55] or distinct approaches from each inside the same intervention session [56].

Developmental Responsive Approach The model's endpoint is occupied by the intervention group known as developmental responsive approaches. These interventions come in two flavors: the developmental, responsive teaching (RT) [57, 58] and individual-difference relationship-based model (DIR) [59]. The RT model has received more in-depth research in the literature than the DIR model. The RT authors recommend several intervention strategies for children, including entering their world; observing their behaviors; mimicking their behaviors and communication patterns; using tone of voice, gestures, and nonverbal cues to enhance communication; engaging in repetitive fun activities; interpreting their conduct as a sign of interest; drawing attention to themselves; and paying attention to their feelings. These strategies help facilitate eye contact, increase intentional communication effectiveness, and use multiple cues to follow the child's attention source.

The RT model and other developmental approaches prioritize building a strong emotional bond between a child and a specialist, as it boosts their motivation for social communication and simplifies their interpretation of social cues. These methods help to shield families from using coercion or manipulation to force a child to exhibit a particular target behavior or from punishing a child for misbehaving. Insofar as they may interfere with the establishment and preservation of a healthy emotional bond between the child and the family, manipulation and punishment methods may have a detrimental emotional impact on the child.

Relationship-based approaches, or RBAs, are used today to support parent-child interactions and child development. Among the emerging practices [60], the RBAs center the intervention on the development and socio-emotional skills of children around their parents. They make an effort to divert attention from the activities that involve direct instruction that experts carry out in clinics or classrooms. Therefore, rather than just teaching kids new words or behaviors, the RBA method seeks to increase the quantity of stimulation children receive by optimizing the quality and effectiveness of the activities parents engage in with their children as part of their everyday routines. The RBA emphasizes the psychosocial effects of parents and caregivers on all children [61], including those with disabilities, and is based on the child developmental parental approach. This participatory method takes into account the child's genetic and biological traits in addition to the caliber of experiences and interactions experienced in regular daily activities [62–64].

Relationship-based techniques encourage parents to be more mindful in their interactions with their young children by focusing on their psychological and social and growth requirements. Research shows that these strategies work well in assisting children who are at risk for language and cognitive development. Scientific studies show these approaches are effective in supporting children with language and cognitive development risks. Studies also reveal that interactive techniques like "Take a turn and wait," "Follow your child's lead," and "Imitate your child" can help parents engage with their kids in a responsive manner. It has been demonstrated that interventions lasting 6 months or longer affect how well children's development is going [65, 66].

Collaborating with Parents A cooperative home-school partnership is crucial for meeting student needs. It benefits students by combining teachers' skills with parents' knowledge. This partnership leads to better academic and social achievement, positive attitudes, higher attendance, and enhanced parent-teacher communication. Parents are the ones who understand their children the best and have a track record of helping them [67].

A collaborative home-school partnership involves a team approach to program planning, clear communication, and understanding of the child's home and school realities.

Build an Informed Partnership Each member of the home-school partnership must have background knowledge before starting the collaboration process in order to be an informed partner. Instructors must possess a broad comprehension of the type of disorder the student has, the student's past, the success of prior interventions, and the student's particular areas of strength and development. It is critical for educators to comprehend the experiences parents have had and to be globally aware of the aspirations, dreams, and goals they have for their kids.

In turn, parents must have a complete awareness of the workings of the educational system, the programs that are offered, and the processes by which decisions about education are made in the school.

Establish a Team Approach The collaborative approach in partnerships is crucial for effectively programming for children with ASD. This involves parents and educators working together, including administrators, special education consultants, teachers, and community resource workers. The school-based team is essential in creating social and academic programs, setting goals, and creating positive behavior plans. They also develop in-services and programs to increase understanding of autism spectrum disorders [65–67].

Plan for Communication Effective communication is crucial for teamwork and maximizing children's potential in school and home settings. Parents and educators should collaborate to create a successful communication strategy, covering regular interactions and handling urgent concerns, beyond traditional interview formats.

Establishing regular meeting times to review IPPs is another step in the process. It is important to think about using a range of communication techniques, such as regular phone calls, notes, letters, journals, daily diaries, and homeschool communication books.

Every day, detailed reports about their children's behavior and learning are frequently requested by parents. While effective collaboration between the home and the school depends on meaningful communication, it is crucial that teachers avoid becoming overburdened by the reporting requirements. Setting daily discussion topics for teachers and parents to work through in order of importance is frequently beneficial. To make sure that the communication protocol still meets the needs of teachers and parents, it should be reviewed on a regular basis.

Teachers should be aware of the kind and quantity of communication that parents need, as well as the fact that parents want to hear good things about their kids. Like all parents, they value heartfelt notes acknowledging their kids' accomplishments and development.

Facilitate Collaborative Relationships Meetings with parents are a common method for teachers to learn about a child, family, and support systems. These meetings help define roles, establish goals, and create strategies for efficient education program execution. It is possible that parents collaborate with other experts who can offer insightful advice on creating behavior plans and educational programs. Find out who parents believe ought to be present at meetings so that they can exchange information and suggest tactics.

Professionals from the school division may also be available to schools, and they can help with planning and collaboration. Here are some ideas for organizing and holding parent-teacher conferences.

I. *Planning the Meeting*

Planning a meeting involves defining the purpose, goals, participants, facilitator, timing, interests, and relationships. It is crucial to consult with school staff and parents, select a facilitator, address interpersonal issues, and develop a neutral agenda.

II. *Conducting a Successful Collaboration Meeting*

Prepare the space and create a welcoming atmosphere for the meeting by ensuring everyone is comfortable and welcome. Allow each participant to share their viewpoint and avoid judgment. Congratulate and offer encouragement, and clarify perspectives when necessary. Treat yourself with consideration and decency, balancing compliments and remarks about difficulties. Give a description of the child's behavior rather than passing judgment. Come up with solutions to both parties' concerns, considering available resources and abilities. At the end of the meeting, summarize points raised and discuss cooperative steps for teachers and parents. Schedule meetings or follow-up communications with specific dates and times.

III. *Follow-Up and Evaluation*

Assess meeting objectives, consider productivity enhancements, and create notes for future meetings [68–70].

1.7 Conclusion

The neurodevelopmental disorder known as ASD is still complicated and varied, with a wide variety of symptoms and behaviors. The increasing number of people with ASD highlights how urgently more research and funding are needed in order to comprehend and treat this disorder. According to epidemiological research, people with ASD are affected by it regardless of their race, ethnicity, or socioeconomic status. ASD has a complex etiology that combines neurological, environmental, and genetic components. Genetic research has revealed several genes linked to ASD, but there is still much to learn about the relationship between these genetic predispositions and environmental factors. Creating focused treatments and preventive measures requires an understanding of these intricate relationships. ASD is characterized by deficits in social communication and engagement, necessitating personalized assessment techniques due to individual differences in these features. These assessments are essential for determining each person's unique strengths and weaknesses, which helps shape customized intervention techniques that can significantly enhance quality of life and cognitive implications. Diverse strategies including educators, therapists, healthcare professionals, and, most importantly, parents and caregivers actively involved are necessary for the successful treatment of ASD. Collaboration between these parties guarantees that therapies are evidence-based, culturally aware, and designed to meet the unique needs of every individual with ASD. Positive outcomes have been demonstrated to be most effectively promoted by parent-mediated treatments and family support services.

In order to effectively address the difficulties of ASD, a coordinated effort combining epidemiological insights, personalized evaluations, and cooperative intervention options is required. To improve our knowledge and management of ASD, further study into the genetic and environmental causes of the illness is essential, as is the creation of novel therapeutic strategies. Working together, researchers, clinicians, educators, and families can create supportive settings that enable persons with ASD to feel better overall and realize their full abilities.

References

1. American Psychiatric Association. DSM-5 Task Force. Diagnostic and statistical manual of mental disorders: DSM-5™. 5th ed. Arlington: American Psychiatric Publishing, Inc; 2013. https://doi.org/10.1176/appi.books.9780890425596.
2. World Health Organization. The ICD-10 classification of mental and behavioural disorders: clinical descriptions and diagnostic guidelines. Geneva: World Health Organization; 1992. https://iris.who.int/handle/10665/37958
3. American Psychiatric Association. Diagnostic and statistical manual of mental disorders: fourth edition, text revision (DSM-IV-TR). 4th ed. Washington, DC: American Psychiatric Association; 2000. https://doi.org/10.1176/appi.books.9780890423349.
4. Vahia VN. Diagnostic and statistical manual of mental disorders 5: a quick glance. Indian J Psychiatry. 2013;55(3):220–3. https://doi.org/10.4103/0019-5545.117131.
5. World Health Organization. International statistical classification of diseases and related health problems (11th Revision). Geneva: World Health Organization; 2018. https://icd.who.int/browse11/l-m/en
6. Regier DA, Kuhl EA, Kupfer DJ. The DSM-5: classification and criteria changes. World Psychiatry. 2013;12(2):92–8. https://doi.org/10.1002/wps.20050.
7. Baxter AJ, Brugha TS, Erskine HE, et al. The epidemiology and global burden of autism spectrum disorders. Psychol Med. 2015;45(3):601–13. https://doi.org/10.1017/S003329171400172X.
8. Baio J, Wiggins L, Christensen DL, et al. Prevalence of autism spectrum disorder among children aged 8 years—autism and developmental disabilities monitoring network, 11 sites, United States, 2014. MMWR Surveill Summ. 2018;67(45):1280. https://doi.org/10.15585/mmwr.ss6706a1.
9. Hodges H, Fealko C, Soares N. Autism spectrum disorder: definition, epidemiology, causes, and clinical evaluation. Transl Pediatr. 2020;9(Suppl 1):S55–65. https://doi.org/10.21037/tp.2019.09.09.
10. Kogan MD, Vladutiu CJ, Schieve LA, et al. The prevalence of parent-reported autism spectrum disorder among US children. Pediatrics. 2018;142(6):e20174161. https://doi.org/10.1542/peds.2017-4161.
11. Baio J, Wiggins L, Christensen DL, et al. Prevalence of autism spectrum disorder among children aged 8 years—autism and developmental disabilities monitoring network, 11 sites, United States, 2014. MMWR Surveill Summ. 2018;67(19):564. https://doi.org/10.15585/mmwr.ss6706a1.
12. Xu G, Strathearn L, Liu B, et al. Prevalence of autism spectrum disorder among US children and adolescents, 2014–2016. JAMA. 2018;319(5):505. https://doi.org/10.1001/jama.2017.17812.
13. Palinkas LA, Mendon SJ, Hamilton AB. Annual review of public health innovations in mixed methods evaluations. Annu Rev Public Health. 2019;40:423–42. https://doi.org/10.1146/annurev-publhealth-040218-044215.
14. Demily C, Poisson A, Peyroux E, et al. Autism spectrum disorder associated with 49, XYYYY: case report and review of the literature. BMC Med Genet. 2017;18(1):9. https://doi.org/10.1186/s12881-017-0371-1.
15. Tartaglia NR, Wilson R, Miller JS, et al. Autism spectrum disorder in males with sex chromosome aneuploidy: XXY/Klinefelter syndrome, XYY, and XXYY. J Dev Behav Pediatr. 2017;38(3):197–207. https://doi.org/10.1097/DBP.0000000000000429.
16. Loomes R, Hull L, Mandy WPL. What is the male-to-female ratio in autism spectrum disorder? A systematic review and meta-analysis. J Am Acad Child Adolesc Psychiatry. 2017;56(6):466–74. https://doi.org/10.1016/j.jaac.2017.03.013.
17. Volkmar F, Siegel M, Woodbury-Smith M, et al. American Academy of Child and Adolescent Psychiatry (AACAP) Committee on Quality Issues (CQI). Practice parameter for the assessment and treatment of children and adolescents with autism spectrum disorder. J Am Acad Child Adolesc Psychiatry. 2014;53(2):237–57. https://doi.org/10.1016/j.jaac.2013.10.013.

18. Bargiela S, Steward R, Mandy W. The experiences of late-diagnosed women with autism spectrum conditions: an investigation of the female autism phenotype. J Autism Dev Disord. 2016;46(10):3281–94. https://doi.org/10.1007/s10803-016-2872-8.
19. Girault JB, Piven J. The neurodevelopment of autism from infancy through toddlerhood. Neuroimaging Clin N Am. 2020;30(1):97–114. https://doi.org/10.1016/j.nic.2019.09.009.
20. Skefos J, Cummings C, Enzer K, et al. Regional alterations in Purkinje cell density in patients with autism. PLoS One. 2014;9(2):e81255. https://doi.org/10.1371/journal.pone.0081255.
21. Stoodley CJ, D'Mello AM, Ellegood J, et al. Altered cerebellar connectivity in autism and cerebellar-mediated rescue of autism-related behaviors in mice. Nat Neurosci. 2017;20(12):1744–51. https://doi.org/10.1038/s41593-017-0004-1.
22. De Rubeis S, He X, Goldberg AP, et al. Synaptic, transcriptional and chromatin genes disrupted in autism. Nature. 2014;515(7526):209–15. https://doi.org/10.1038/nature13772.
23. Shen MD, Kim SH, McKinstry RC, et al. Increased extra-axial cerebrospinal fluid in high-risk infants who later develop autism. Biol Psychiatry. 2017;82(3):186–93. https://doi.org/10.1016/j.biopsych.2017.02.1095.
24. Hazlett HC, Gu H, Munsell BC, et al. Early brain development in infants at high risk for autism spectrum disorder. Nature. 2017;542(7641):348–51. https://doi.org/10.1038/nature21369.
25. Siddiqui RW, Siddiqui TW, Siddiqui SW. Vitamin D and autism spectrum disorder: an intriguing association. Ann Child Neurol. 2024;32(2):83–91.
26. Panda PK. Current consensus on clinical features, pathogenesis, diagnosis and management of autism spectrum disorder in children: a brief review. Pediatr Rev Int J Pediatr Res. 2019;6(3):144–9. https://pediatrics.medresearch.in/index.php/ijpr/article/view/478
27. Surén P, Roth C, Bresnahan M, et al. Association between maternal use of folic acid supplements and risk of autism spectrum disorders in children. JAMA. 2013;309(6):570–7. https://doi.org/10.1001/jama.2012.155925.
28. Frye RE, Rossignol DA, Scahill L, McDougle CJ, Huberman H, Quadros EV. Treatment of folate metabolism abnormalities in autism spectrum disorder. Semin Pediatr Neurol. 2020;35:100835. https://doi.org/10.1016/j.spen.2020.100835.
29. Aldera H, Hilabi A, Elzahrani MR, et al. Do parental comorbidities affect the severity of autism spectrum disorder? Cureus. 2022;14(12):e32702. https://doi.org/10.7759/cureus.32702.
30. Jones KL, Van de Water J. Maternal autoantibody related autism: mechanisms and pathways. Mol Psychiatry. 2019;24(2):252–65. https://doi.org/10.1038/s41380-018-0099-0.
31. Malkova NV, Yu CZ, Hsiao EY, et al. Maternal immune activation yields offspring displaying mouse versions of the three core symptoms of autism. Brain Behav Immun. 2012;26(4):607–16. https://doi.org/10.1016/j.bbi.2012.01.011.
32. Schieve LA, Tian LH, Drews-Botsch C, et al. Autism spectrum disorder and birth spacing: findings from the study to explore early development (SEED). Autism Res. 2018;11(1):81–94. https://doi.org/10.1002/aur.1887.
33. Laverty C, Surtees A, O'Sullivan R, et al. The prevalence and profile of autism in individuals born preterm: a systematic review and meta-analysis. J Neurodev Disord. 2021;13(1):41. https://doi.org/10.1186/s11689-021-09382-1.
34. Girolamo T, Shen L, Monroe Gulick A, Rice ML, Eigsti IM. Studies assessing domains pertaining to structural language in autism vary in reporting practices and approaches to assessment: a systematic review. Autism. 2024;28(7):1602–21. https://doi.org/10.1177/13623613231216155.
35. Xie F, Pascual E, Oakley T. Functional echolalia in autism speech: verbal formulae and repeated prior utterances as communicative and cognitive strategies. Front Psychol. 2023;14:1010615. https://doi.org/10.3389/fpsyg.2023.1010615.
36. Yang K. Where there is an echo, there is an intention: understanding the echolalia phenomenon of children with epilepsy and autism. Commun Disord Q. 2023;45(1):42–52. https://doi.org/10.1177/15257401221132763.
37. Vives-Vilarroig J, Ruiz-Bernardo P, García-Gómez A. Sensory integration and its importance in learning for children with autism spectrum disorder. Cadernos Brasileiros de Terapia Ocupacional. 2022;30:e2988. https://doi.org/10.1590/2526-8910.ctoAR22662988.

38. Parham LD and Mailloux Z (2010) Sensory Integration. In Smith JC. and O'Brain JC., (eds.) Occupational Therapy for Children, 6th Edition, Elsevier Inc., 349–355.
39. Gray C, Garand JD. Social stories: improving responses of students with autism and with accurate social information. Focus Autism Other Dev Disabl. 1993;8(1):1–10.
40. Friedlander R. Outcomes in neurodevelopmental and genetic disorders. Can Child Adolesc Psychiatr Rev. 2005 May;14(2):57.
41. Bristol MM, Cohen DJ, Costello EJ, et al. State of the science in autism: report to the National Institutes Health. J Autism Dev Disord. 1996;26(2):121–54. https://doi.org/10.1007/BF02172002.
42. Kana RK, Keller TA, Cherkassky VL, Minshew NJ, Just MA. Sentence comprehension in autism: thinking in pictures with decreased functional connectivity. Brain. 2006;129(9):2484–93. https://doi.org/10.1093/brain/awl164.
43. Alberta Learning. Teaching students with autism: a guide for educators. Alberta: Alberta Learning Cataloguing in Publication; 1999.
44. Koegel RL, Vernon TW, Koegel LK. Improving social initiations in young children with autism using reinforcers with embedded social interactions. J Autism Dev Disord. 2009;39(9):1240–51. https://doi.org/10.1007/s10803-009-0732-5.
45. DeFilippis M, Wagner KD. Treatment of autism spectrum disorder in children and adolescents. Psychopharmacol Bull. 2016;46(2):18–41.
46. Ellerbeck K, Smith C, Courtemanche A. Care of children with autism spectrum disorder. Prim Care. 2015;42(1):85–98. https://doi.org/10.1016/j.pop.2014.09.004.
47. Robins DL, Casagrande K, Barton M, Chen CM, Dumont-Mathieu T, Fein D. Validation of the modified checklist for autism in toddlers, revised with follow-up (M-CHAT-R/F). Pediatrics. 2014;133(1):37–45. https://doi.org/10.1542/peds.2013-1813.
48. Ehlers S, Gillberg C, Wing L. A screening questionnaire for Asperger syndrome and other high-functioning autism spectrum disorders in school age children. J Autism Dev Disord. 1999;29(2):129–41. https://doi.org/10.1023/a:1023040610384.
49. Huerta M, Lord C. Diagnostic evaluation of autism spectrum disorders. Pediatr Clin N Am. 2012;59(1):103. https://doi.org/10.1016/j.pcl.2011.10.018.
50. Hyman SL, Levy SE, Myers SM, et al. Identification, evaluation, and management of children with autism spectrum disorder. Pediatrics. 2020;145(1):e20193447. https://doi.org/10.1542/peds.2019-3447.
51. Clark B, Vandermeer B, Simonetti A, Buka I. Is lead a concern in Canadian autistic children? Paediatr Child Health. 2010;15(1):17–22.
52. Dosman C, Witmans M, Zwaigenbaum L. Iron's role in paediatric restless legs syndrome—a review. Paediatr Child Health. 2012;17(4):193–7. https://doi.org/10.1093/pch/17.4.193.
53. Filipek PA, Accardo PJ, Ashwal S, Baranek GT, Cook EH, Dawson G, et al. Practice parameter: Screening and diagnosis of autism. Report of the Quality Standards Subcommittee of the American Academy of Neurology and the Child Neurology Society. Neurology. 2000;55(4):468–79. https://doi.org/10.1212/WNL.55.4.468.
54. Harris C, Card B. A pilot study to evaluate nutritional influences on gastrointestinal symptoms and behavior patterns in children with autism spectrum disorder. Complement Ther Med. 2012;20(6):437–40. https://doi.org/10.1016/j.ctim.2012.08.004.
55. Reebye P. Social & communication development in autism spectrum disorders: early identification, diagnosis, & intervention. J Can Acad Child Adolesc Psychiatry. 2008;17(2):93–5.
56. Buffington DM, Krantz PJ, McClannahan LE, Poulson CL. Procedures for teaching appropriate gestural communication skills to children with autism. J Autism Dev Disord. 1998;28(6):535–45. https://doi.org/10.1023/a:1026056229214.
57. Roxburgh CA, Carbone VJ. The effect of varying teacher presentation rates on responding during discrete trial training for two children with autism. Behav Modif. 2013;37(3):298–323. https://doi.org/10.1177/0145445512463046.
58. Jobin A. Varied treatment response in young children with autism: a relative comparison of structured and naturalistic behavioral approaches. Autism. 2020;24(2):338–51. https://doi.org/10.1177/1362361319859726.

59. Prelock PJ, Nelson NW. Language and communication in autism: an integrated view. Pediatr Clin N Am. 2012;59(1):129–45. https://doi.org/10.1016/j.pcl.2011.10.008.
60. Yoder PJ, Warren SF. Effects of prelinguistic milieu teaching and parent responsivity education on dyads involving children with intellectual disabilities. J Speech Lang Hear Res. 2002;45(6):1158–74. https://doi.org/10.1044/1092-4388(2002/094).
61. Whalen C, Schreibman L. Joint attention training for children with autism using behavior modification procedures. J Child Psychol Psychiatry. 2003;44(3):456–68. https://doi.org/10.1111/1469-7610.00135.
62. Toth K, Munson J, Meltzoff AN, Dawson G. Early predictors of communication development in young children with autism spectrum disorder: joint attention, imitation, and toy play. J Autism Dev Disord. 2006;36(8):993–1005. https://doi.org/10.1007/s10803-006-0137-7.
63. Mahoney G, Perales F, Wiggers B, Herman B. Responsive teaching: early intervention for children with Down syndrome and other disabilities. Downs Syndr Res Pract. 2006;11(1):18–28. https://doi.org/10.3104/perspectives.311.
64. Selimoglu OG, Emel Sardohan Yildirim A. General overview of evidence-based practices for autism spectrum disorder. Eur J Spec Educ Res. 2019;5(2):29–42. https://doi.org/10.5281/zenodo.3546443.
65. Shumway S, Wetherby AM. Communicative acts of children with autism spectrum disorders in the second year of life. J Speech Lang Hear Res. 2009;52(5):1139–56. https://doi.org/10.1044/1092-4388(2009/07-0280).
66. Odom SL, Hall LJ, Morin KL. Educational interventions for children and youth with autism: a 40-year perspective. J Autism Dev Disord. 2021;51(12):4354–69. https://doi.org/10.1007/s10803-021-04990-1.
67. Franz L, Goodwin CD, Rieder A, Matheis M, Damiano DL. Early intervention for very young children with or at high likelihood for autism spectrum disorder: an overview of reviews. Dev Med Child Neurol. 2022;64(9):1063–76. https://doi.org/10.1111/dmcn.15258.
68. Bowman KS, Suarez VD, Weiss MJ. Standards for interprofessional collaboration in the treatment of individuals with autism. Behav Anal Pract. 2021;14(4):1191–208. https://doi.org/10.1007/s40617-021-00560-0.
69. Yeom JS, Kim J. Effectiveness of online responsive teaching in young children with developmental disabilities: a pilot study. Clin Exp Pediatr. 2024;67(6):303–11.
70. Cassel TD, Messinger DS, Ibanez LV. Early social and emotional communication in the infant siblings of children with autism spectrum disorders: an examination of the broad phenotype. J Autism Dev Disord. 2007;37(1):122–32. https://doi.org/10.1007/s10803-006-0337-1.

The Comprehensive Assessment of the Autistic Children

2

Neha Sharma

Learning Objectives

- After studying this chapter, you should be able to distinctly explain:
- The importance of early identification of ASD
- Important tools used for screening and assessing ASD at an early stage
- Assessment areas to be focused in children with ASD
- Detailed cognitive, sensory perceptual, motor, and language assessment in children with ASD

2.1 Brief Introduction

ASD is a neurodevelopmental condition which includes challenges faced by children in communicating with other individuals and social interaction and showing repetitive behaviours and restricted interest [1]. Physiotherapy assessment in these children helps in assessing the physical and social challenges faced by children with ASD and also helps in overcoming these challenges and improving their quality of life.

2.2 Identification of ASD at an Early Stage

A clinical implication of early detection of ASD is the ability to refer young children, between the ages of 2 and 5 years, for assessment and treatment services as early as possible [1, 2]. Early detection can help therapists to develop an

N. Sharma (✉)
Maharishi Markandeshwar Institute of Physiotherapy and Rehabilitation, Maharishi Markandeshwar (Deemed to be University), Mullana - Ambala, Haryana, India
e-mail: neha_sharma@mmumullana.org

M. Goyal, K. Goyal (eds.), *Rehabilitation Approach in Autism*,
https://doi.org/10.1007/978-981-96-4162-8_2

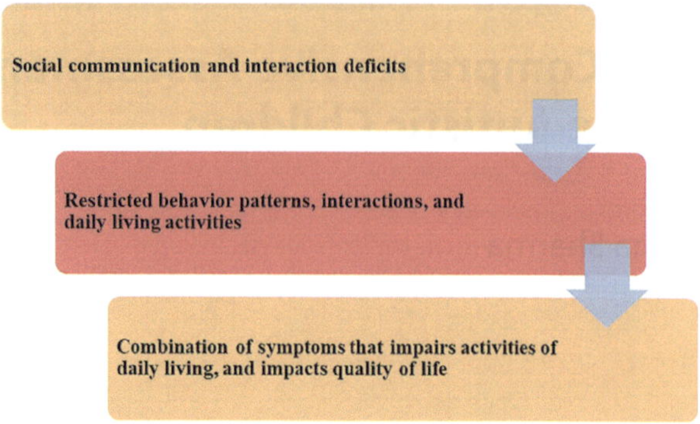

Fig. 2.1 Common symptoms in children with ASD

intervention protocol addressing issues in young children with communication, movement skills, and social interaction. Early detection of ASD at a young age also helps in improving the quality of life of the child with ASD [3], displayed in Fig. 2.1.

2.2.1 Benefits of Early Detection [4]

1. Early detection helps in improving cognitive, language, and behavioural functions in children with ASD.
2. Early detection of ASD promotes early intervention of ASD which improves developmental outcomes and quality of life in children with ASD.
3. Parents who support early detection and management of ASD have a lower level of anxiety and stress in their lives.
4. During adulthood, children with ASD have better social outcomes and good independence, if early detection and management was followed.
5. It also reduces long-term intensive intervention and social education services which also promotes cost savings.

2.3 Screening and Diagnostic Methods for ASD

While autism spectrum disorder can occasionally be diagnosed at 18 months or younger, a diagnosis made by a qualified expert is considered reliable at the time when the child is 2 or more than 2 years of age [5]. It is advised by the National Center on Birth Defects and Developmental Disabilities (NCBDD) that children must have proper screening for ASD at 9, 18, 24, and 30 months of age. The American Academy of Pediatrics (AAP) also suggests screening of children between 1 and 2 years of age [6].

The diagnosis of ASD depends on an understanding of children's developmental milestones. The developmental milestones is useful for parents and paediatric medical professionals to monitor the growth parameters of a child from birth to age 5 [6]. These benchmarks include language and communication, social and emotional abilities, and both mental and physical development. These developmental milestones provide guidance of growth patterns of a child and also helps in identifying developmental changes and concerns [3], as displayed in Fig. 2.2.

Screening of child with ASD at an early phase do not provide the definite diagnosis, but may indicate the need of developmental evaluation [2]. Paediatricians and paediatric physiotherapists are experts which help in determining whether the child fits the diagnostic criteria for ASD throughout this phase of development [4].

2.3.1 Common Screening and Diagnostic Tools Used for Assessing ASD

1. *Modified Checklist for Autism in Toddlers with Follow-Up (M-CHAT-R/F)* [7]

This checklist, developed by Robins, Fein, and Barton in 2009, is a parent reporting screening tool that assesses and screens children between the age of 16 and 30 months [7]. Total scoring consists of 20 scores, with a total of 20 items, 1 score for each item. M-CHAT-R/F can be used by professionals as well as by parents. Cronbach's alpha for this scale is 0.79. The scale has sensitivity of 0.83 and specificity of 0.99 [8]. The scale is used at the time of baseline or first assessment and also can be used at the time of follow-up [7].

2. *The Ages and Stages Questionnaires (ASQ)* [9]

This questionnaire is a parent-reported tool, which screens children between 2 and 66 months of age. The total duration required for assessing the child with ASD with ASQ is between 18 and 20 min [9]. ASQ consists a total of 30 items, which focuses on five areas: personal social, fine motor, gross motor, problem-solving, and communication [10]. This questionnaire has 92% test-retest reliability, 87.4% sensitivity, and 95.7% specificity [9].

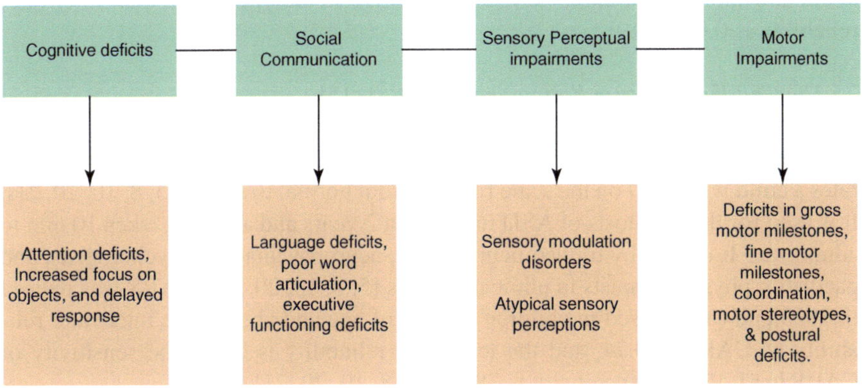

Fig. 2.2 Impairments in ASD

3. *The Screening Tool for Autism in Toddlers (STAT)* [11]

The STAT is used for assessing children between 24 and 36 months of age [12]. STAT was developed by Stone et al. in 2000 and then revised in 2004 and 2008 [11]. It comprises a total of 12 activities which assess key social-communicative behaviours in terms of play, communication, and motor activity. It also takes a total of 20 min to administer. The inter-rater reliability for STAT is $\kappa = 1.00$ and test-retest reliability $\kappa = .90$. It has high sensitivity of 92% and specificity of 85% [11].

4. *The Social Communication Questionnaire (SCQ)* [13]

The SCQ is a parent-completed tool used for assessing symptoms of ASD in children with and more than 4 years of age [14]. It is a 40-item semistructured parent interview based on social interactions, language development, and symptoms of suspected ASD [13]. The SCQ has Cronbach's alpha coefficient of 0.95. The validity of SCQ is 0.98. It has sensitivity of 85% and specificity of 25.7%.

5. *The Parent's Evaluation of Developmental Status (PEDS)* [15]

PEDS is also a parent-reported screening tool, used for assessing ASD in children from birth to 8 years of age. It was developed by Frances Page Glascoe [8, 10, 16–18]. It consists of a ten-item questionnaire based on language, motor development, behaviour, and learning. PEDS has a test-retest reliability of 0.98 and 0.99 and inter-rater reliability of 0.82 and 0.96. The sensitivity of PEDS is 57.1%, and specificity is 97.6% [8, 10, 16–18].

6. *Autism Diagnosis Interview-Revised (ADI-R)* [15]

This interview guide is a parent-completed interview guide which is used to screen children with ASD at least with the mental age of 2 years [8, 19]. ADI-R takes 2 h to administer, having a total of 93 items in three domains of functioning: communication, social interaction, and repetitive stereotyped behaviour. Inter-rater reliability is 0.83 [8, 19].

7. *The Childhood Autism Rating Scale (CARS)* [15]

CARS rates behaviour in a child with ASD. It consists a total of 15 items that rates a child with ASD on the scale ranging from normal to severe [5, 8, 10, 20, 21]. It is used for the diagnosis of ASD in children 2 years and above. It takes 10 min to administer. It is used by trained neurologists, neuropsychiatrist, psychologists, and paediatric physiotherapists in clinical settings [5, 8, 10, 20, 21]. CARS was revised in 2003 and can be used in children 6 years of age and above. The inter-rater reliability of CARS is 0.74, and the test-retest reliability is 0.81. The sensitivity of CARS is 81.4%, and specificity is 78.4% [5, 8, 10, 20, 21].

8. *The Autism Diagnostic Observation Schedule-Generic (ADOS-G)*

The ADOS-G tool used for assessing a child's interaction and communication skills to diagnose ASD [5, 12, 14, 19]. The observation tool consists of 30-min modules. Each module is designed accordingly to assess different patterns of interaction and communication. ADOS was first introduced in the 1980s as a method of observations of social behaviour, communication, and play in children suspected with ASD. The tool can be used after 1 year of age [5, 12, 14, 19]. It takes 40–60 min to administer. ADOS-G has >97% sensitivity and >77% specificity [5, 12, 14, 19].

9. *Indian Scale for Assessment of Autism (ISAA)*

This tool consists of 40 items rated on a 5-point scale ranging from 1 (never) to 5 (always) [20]. The 40 items are divided under six domains as given below: emotional responsiveness, social relationship and reciprocity, and communication. Among children with an age group between 2 and 9 years, ISAA has a sensitivity of 93.3% and specificity of 97.4%. The test-retest reliability is 0.76, and inter-rater reliability is 0.87 [20].

2.4 ASD and Its Impairments [3, 15]

2.4.1 Cognitive Deficits

Infants and children with ASD having attention deficits consist of delayed or lack of response and difficulty in disengaging attention [19].

2.4.2 Social Communication

Social communication is the hallmark of diagnosing ASD. In this, infants and children with ASD have impaired pragmatic language such as poor use of nonverbal cues, i.e., eye gaze, facial expressions, poor word articulation, and immediate or delayed imitation of words [13, 14, 22].

They also have delay in executive functions such as problem-solving, which consists of planning, response, inhibition, organized search, and impulse control. Children with ASD show repetitive behaviours and difficulties in inhibiting functional movements.

2.4.3 Sensory Perceptual Impairments

Children with ASD show sensory perceptual impairments in two ways; they have either sensory modulation disorders or atypical sensory processing.

2.4.3.1 Sensory Modulation Disorders

In this, the child faces difficulties in organizing and responding to different sensory stimuli such as touch, smell, auditory, visual, vestibular, and kinaesthetic [10]. Here, the child can be "under-responsive", i.e. slow to response or fails to react to name or other sensory stimuli. The child with ASD also shows "over-responsiveness". The child with ASD may show exaggerated behaviour towards different sensory stimuli, such as covering both the eyes with hands when seeing bright colours [8].

2.4.3.2 Atypical Sensory Perceptions

Infants and children with ASD show increase in local processing. They are not able to understand and identify the relationships between more than one sensory stimulus.

2.4.4 Motor Delays or Impairments

Infants and children with ASD show delay in growth parameters (developmental milestones).

2.4.5 Gross Motor Coordination

In infants, toddlers, and preschoolers, gross motor delays are evident in supine, prone, and sitting positions and also in walking. Gross motor milestones are affected in infants with ASD. In children above 6 years of age, poor coordination in the upper and lower limbs are evident along with impaired visuomotor coordination.

2.4.6 Fine Motor Coordination

In infants, toddlers, and preschoolers, delay in reaching and grasping is present. In preschoolers and middle-school-going children, poor manual dexterity is observed.

2.4.7 Motor Stereotypes

Banging of objects, and repetitive behaviour patterns are observed in children less than 6 years of age, while, in middle-school-going children, repetitive behaviours such as jumping, bouncing, arm flapping, and body rocking are evident [18].

2.4.8 Postural Deficits

Delays in rolling, sitting, and presence of unusual postures are evident in infants and preschoolers [5]. Feed-forward and feedback control is affected in middle-school-going children having ASD.

2.5 ASD Examination Procedure

2.5.1 History Taking

Age, growth parameters, and detailed history will be asked by the parents or caregivers.

2.5.2 Behavioural Observations

Child is playing and using non-verbal communication (use of facial gestures, communication skills, and interaction) or verbal communication (asking no. of words used in a phrase, using language milestones based on developmental status). The therapist should observe for repetitive behaviours in a child with ASD, gross motor skills, fine motor skills, stair climbing, and kicking a ball.

2.5.3 Cognitive Assessment

Cognitive assessment in children with ASD should be done with the Stanford-Binet Intelligence Test (SBIT), which is used for assessing the intelligent quotient (IQ) of children between 2 and 17 years of age. Another scale, the Wechsler Intelligence Scale for children between 6 and 16 years of age can also be used for IQ testing.

Cognitive components can also be assessed by using the Montreal Cognitive Assessment scale in children above 14 years of age. Components of cognition can be assessed in children with ASD by asking the name of colours, flowers, and animals, copying a rectangle or square (visuospatial orientation), sharing memories, attention, and language.

2.5.4 Sensory Perceptual Assessment

A child with ASD can show hypo- or hyperresponsiveness to particular sensory stimuli (tactile, visual, olfactory, auditory, visual, vestibular, and proprioceptive).

Parent or caregiver questionnaires can also be used for assessing sensory-perceptual impairments with the infant and toddler sensory profile in children up to 3 years of age and also with the short sensory profile which can be used for assessing sensory perceptual impairments in children between 3 and 14 years of age.

2.5.5 Motor Examination

Developmental milestones should be assessed, firstly, in children with ASD. Specific scales used for motor functions in a child with ASD are the Alberta Infant Motor Scale (AIMS) for infants and toddlers, Movement Assessment Battery for Children

(MABC), Peabody Developmental Motor Scales (PDMS-2), and Bruininks-Oseretsky Test of Motor Proficiency (BOTMP).

2.5.6 Strength and Tone Assessment

For assessing strength in children with ASD, manual muscle testing can be used. Muscle tone can be examined with the Modified Ashworth Scale (MAS) and also with the Tardieu Scale.

2.5.7 Functional Assessment

Functional assessment in children with ASD can be done by the Paediatric Evaluation of Disability Inventory (PEDI), for assessing functional activities, social interaction, and quality of life. It is used in children between 6 months and 7 years of age.

2.5.8 Types of Developmental Screening Tools

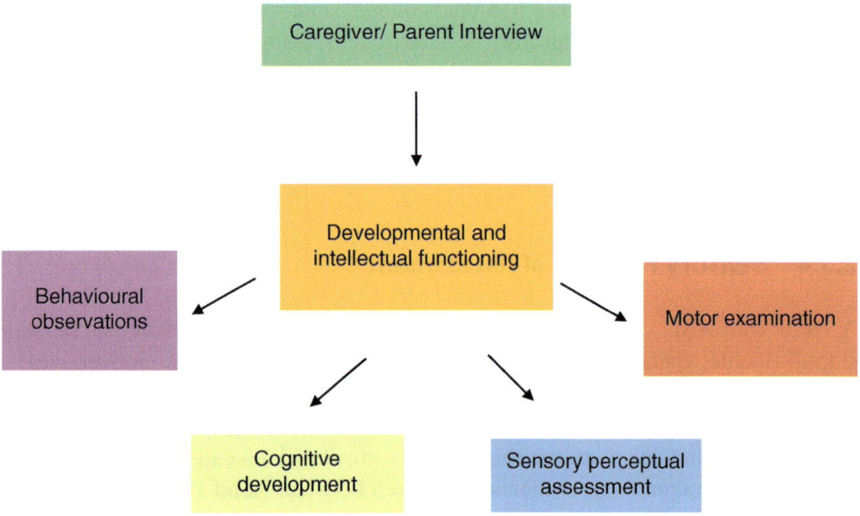

Fig. 2.3 Areas to be focused during evaluation of autism

2.6 Future Methods for Screening and Diagnosis

Creating biomarkers in combination with the screening techniques helps in increasing the diagnostic professional's effectiveness [11]. Biomarkers which help in stratifying risks, are used for secondary screening, and confirm the behavioural observations which can be used for diagnosis could greatly improve the quality of ASD assessment and management in clinical settings. Categories of Biomarkers which can be used for screening and stratifying risks and diagnosing and treating children with ASD are prenatal biomarkers which help in identifying high-risk pregnancies. Presymptomatic biomarkers help in identifying fragile infants who require diagnostic testing and treatment for ASD [11] as displayed in Fig. 2.3. Diagnostic biomarkers are those that confirm the diagnosis after the confirmation of the symptoms of ASD. Subgrouping types of biomarkers are those which can be used to diagnose children with ASD with different ASD types. Biomarkers also help in treating children with ASD by selecting the most optimal treatment or therapy by predicting the response to the treatment [2].

2.7 Physiotherapy Assessment Format for Children with ASD

Demographic details of the child
Name:
Age:
Gender:
Date of birth:
Hand dominance:
Parent details:
Caregiver information (if any):

Growth parameters
Weight:
Height:
Body mass index:
Head circumference:

Chief complaints (by parents or caregiver):

History:
History of present illness:

History of past illness:

Family history:

Maternal history prenatal, antenatal, and postnatal history:

Developmental History

Developmental milestones

Gross motor milestones	Appeared when:
Neck holding	
Rolling	
Sitting with support	
Sitting without support	
Standing with support	
Creeping	
Crawling	
Standing without support	
Walking	
Creeping upstairs	
Running	
Walking stairs up and down	
Ridding tricycle	
Fine motor milestones	
Bidextrous reach	
Unidextrous reach	
Pincer grasp	
Scribbling	
Copies circle	
Copies triangle	
Copies cross	
Language milestones	
Alerts to sound	
Cooing	
Laugh	
Monosyllables	
Bisyllables	
2 words with meaning	
2–3 words sentence formation	
Asking questions	
Identifying and telling gender	
Singing rhyme, song, or poem	
Telling stories	
Asking meaning of words	
Social milestones	
Social smile	
Recognizing mother	
Waves	
Responding when talking	

(continued)

Jargon	
Asking for food and drinks	
Sharing toys	
Playing in groups	
Self-grooming and helping others	

ASD assessment methods:

1. Direct observation of child
2. Manual testing
3. Interview of parent/caregiver

1. ***Direct observation of child***

Physiotherapist should focus on child's particular aspects:
Interaction with parents or caregivers
Interest in playing and toys
Communication with parents, caregivers, or any other children
Comfort level of child
Repetitive agitated or inappropriate behaviour

Motor stereotype observation

Type of activities	Yes	No
Banging of objects or toys		
Chewing shirt sleeves		
Jumping		
Bouncing		
Arm flapping		
Body rocking		

Behavioural observations

Behavioural cues	Yes	No
Poor eye contact		
Social smile		
Using facial gestures for communication		
Interaction with parents, caregivers		
Interaction with other kids		
Following commands		
Shows attentiveness to particular activity		
Enjoys playing in a group		
Prefer playing alone		

(continued)

Behavioural cues	Yes	No
Getting excited or distressed without any reason		
Smile to self without any reason		
Showing anxiety or fear without any reason		
Fear of falling		
Repeating same word or statement		
Showing inappropriate or aggressive behaviour such as kicking or hitting toys or objects		
Use jargon or inappropriate words while communicating		
Yelling or screaming		

2. ***Manual testing***

Muscle strength: By manual muscle testing

Muscle tone: Through modified Ashworth Scale and Tardieu Scale

Functional assessment: PEDI (for 6 months to 7 years of age)
The school function assessment (for children up to 3 years of age)

Motor examination: Can be done through these scales according to the age of the child mentioned in detail above, AMIS, MABC, PDMS-2, and BOTMP.

Sensory perceptual assessment:

Sensory stimuli	Hyporesponsiveness	Hyperresponsiveness
Tactile		
Visual		
Olfactory		
Auditory		
Visual		
Vestibular		
Proprioceptive		

Sensory examination
Pain
Touch
Temperature
Vibration
Position sense
Movement sense

Stereognosis
Barognosis
Tactile localisation
Two-point discrimination
Graphesthesia

Posture examination:

Cognitive assessment

Copy triangle:

Orientation (for children above 5 years of age):

Date:	Day:	Month:	Year:	Time:

Self-orientation and memory (for children above 5 years of age):

Colour of dress:	Mother name:	Father name:	Total days in a week:	Total months in a year:

Coordination assessment:
Fine motor coordination:
Finger to finger:
Finger to nose:
Making circles in the air:
Eye-hand coordination:

Gross motor coordination:
Tandem stand:
Tandem walking:
Waking in a circle:

Gait examination:

Pain examination: The FLACC (face, legs, activity, cry, consolability) scale between 2 months and 7 years of age. Numerical pain rating scale (NPRS) can be used for children more than 7 years of age.

3. **Interview of parent or caregiver**

The interview questions must be asked from the parents or caregivers of the child.

The following questions can be asked by the parents or caregivers.

Questions	Yes	No
1. Responds to loud noise		
2. Covers ears to protect them from sound		
3. Covers eyes to protect them from light		
4. Enjoys bright colours		
5. Cannot lift heavy objects		
6. Enjoys jumping		
7. Difficulty in paying attention		
8. Does not respond to his/her own name		
9. Switches from one activity to another immediately without completing previous task		
10. Feels that he/she falls down when feet leave the ground, e.g. in sitting or lying		
11. Fear of heights		
12. Enjoys solid foods		
13. Only eats liquid food		
14. Avoids food smell or taste		
15. Gets aggressive or emotional when provided with tactile or touch stimuli		
16. Avoids going barefoot on grass		
17. Enjoys spinning on marry-go round and rolling		
18. Enjoys cuddling or covered in a blanket		
19. Gets aggressive or emotional in lifts (vertical direction)		
20. Communicates through facial gestures instead of talking		
21. Interacts with parents or caregivers		
22. Interacts with siblings or other peer kids		
23. Loves to share toys or making friends		
24. Enjoys playing alone		
25. Participates in school activities		
26. Enjoys self-talking		
27. Shows fear to dangers		
28. Drags or pulls parents hand to get specific object		
29. Repeats same word or sentence		
30. Produces bizarre or unwanted sounds		
31. Gets attached to inanimate objects such as toys, blanket, etc.		
32. Shows tantrums (e.g. yelling, hand banging, or screaming)		
33. Biting or hitting, causes self-injurious behaviour		
34. Pain insensitivity		
35. Having interest in solving mathematical problems and artistic work		

Investigation reports (if any)
Radiological investigations (CT scan/MRI):
Biochemistry (blood serum/CSF examination):
Electrophysiological investigations (EEG/EMG/NCV):
Medical intervention (drugs prescribed by concerned physician):
Differential diagnosis:
Problem list (based on ICF pattern):

33. Biting or hitting, causes self injurious behaviour?		
34. Pain insensitivity?		
35. Having interest in solving mathe matical problems, and artistic work?		

Provisional diagnosis:
Physiotherapy intervention goals:

Body function and structure	**Activity**	**Participation**

Physiotherapy management:

Short term goals	**Long term goals**

2.8 Summary

Autism spectrum disorder (ASD) is a neurodevelopmental disorder which includes neurodevelopmental issues such as communicating with other individuals along with social interaction. The diagnosis of ASD depends on an understanding of children's developmental milestones, social behaviour, and interaction with peer and parents. The developmental milestones is useful for parents and healthcare professionals to monitor a child's growth from birth to age 5. Early screening with the help of diagnostic tools helps in managing a child with ASD at an early age. Physiotherapy assessment plays a major role in assessing the child's functional and mental status. Keen observation of the child, screening and diagnostic tools, and parent-reported observations help in diagnosing children with ASD. Physiotherapy given as early care helps in improving the functional and mental status of a child with ASD. It also helps in improving the child's social and communication skills as early as possible. It helps in overcoming functional and social challenges and also the quality of life in children with ASD.

References

1. Katakis P, Lockwood G, Jeanne E, Shermina W, Ellie S, Mirzaei V, et al. Diagnostic assessment of autism in children using telehealth in a global context: a systematic review. Rev J Autism Dev Disord [Internet]. 2023;0123456789. https://doi.org/10.1007/s40489-023-00408-z.
2. Armstrong K, Duvall SW. Introductory editorial to the special issue: assessment and diagnosis of autism spectrum disorder (ASD) and related clinical decision making in neuropsychological practice Introductory editorial to the special issue: assessment Clin Neuropsychol [Internet]. 2022;36(5):851–5. https://doi.org/10.1080/13854046.2022.2085629.
3. Mccarty P, Frye RE. Early detection and diagnosis of autism spectrum disorder: why is it so difficult? Semin Pediatr Neurol [Internet]. 2020;35:100831. https://doi.org/10.1016/j.spen.2020.100831.
4. Bishop SL, Lord C. Commentary: best practices and processes for assessment of autism spectrum disorder—the intended role of standardized diagnostic instruments. 2023;5:834–8.
5. Guivarch J, Murdymootoo V, Elissalde S, Salle- X, Tardieu S, Jouve E. Impact of an implicit social skills training group in children with autism spectrum disorder without intellectual disability: a before-and-after study. PLoS One. 2017:1–18.
6. Okoye C, Obialo-ibeawuchi CM, Obajeun OA, Sarwar S, Tawfik C, Waleed MS, et al. Early diagnosis of autism spectrum disorder: a review and analysis of the risks and benefits. 2023;15(8).
7. Oner O, Munir KM. Modified checklist for Autism in Toddlers Revised (MCHAT-R/F) in an urban metropolitan sample of young children in Turkey. J Autism Dev Disord. 2021;50(9):3312–9.
8. Trial P, Cheng Y, Lin Y, Wu Y, Tsai Y. Children with autism spectrum disorder. Nutrients. 2019:1–16.
9. Lockhart M, Chaux R, Chevin M, Celle M, Raia-barjat T, Patural H, et al. Classification performance of the ages and stages questionnaire: influence of maternal education level. 2023. p. 1–9.
10. Toizumi M, Thi G, Nguyen H, Motomura H, Nguyen TH. Sensory defects and developmental delay among children with congenital rubella syndrome. Nat Publ Gr [Internet]. 2017:1–11. https://doi.org/10.1038/srep46483.

11. Dick CC, Attar SM, Ibanez LV, Stone WL. The implementation of the screening tool for autism in toddlers in part C early intervention programs: an 18-month follow-up. Autism. 2024;27(1):173–87.
12. Attar SM, Ibanez LV, Stone WL. Separate scoring algorithms for specific identification priorities optimize the screening properties of the STAT. Autism Res. 2023;15(11):2069–80.
13. Marvin AR, Marvin DJ, Lipkin PH, Law JK. Analysis of Social Communication Questionnaire (SCQ) screening for children less than age 4. 2017. p. 137–44.
14. Levin AR, Nelson CA. Social communication difficulties and autism in previously institutionalized children. J Am Acad Child Adolesc Psychiatry. 2016;54(2):1–17.
15. Yu Y, Ozonoff S, Miller M. Assessment of autism spectrum disorder. 2024.
16. Steinman KJ, Stone WL, Ibañez LV, Attar SM. Reducing barriers to autism screening in community primary care: a pragmatic trial using web-based screening. Acad Pediatr. 2023;22(2):263–70.
17. Ibañez LV, van der Stoep A, Myers K, Zhou C, Dorsey S, Steinman KJ, et al. Promoting early autism detection and intervention in underserved communities: study protocol for a pragmatic trial using a stepped-wedge design. BMC Psychiatry. 2019:1–13.
18. Howard JS, Stanislaw H, Green G, Sparkman CR, Cohen HG. Research in developmental disabilities comparison of behavior analytic and eclectic early interventions for young children with autism after three years. Res Dev Disabil [Internet]. 2014;35(12):3326–44. https://doi.org/10.1016/j.ridd.2014.08.021.
19. Wood JJ, Ehrenreich-may J, Alessandri M, Fujii C, Renno P, Laugeson E, et al. Cognitive behavioral therapy for early adolescents with autism spectrum disorders and clinical anxiety: a randomized, controlled behaviour. Behav Ther. 2016;46(1):7–19.
20. Chakraborty S, Id TB, Antony N, Roy A, Shriharsh V, Sahay A, et al. Comparing the Indian Autism Screening Questionnaire (IASQ) and the Indian Scale for Assessment of Autism (ISAA) with the Childhood Autism Rating Scale—Second Edition (CARS2) in Indian settings. PLoS One [Internet]. 2022;2:1–15. https://doi.org/10.1371/journal.pone.0273780.
21. Al-ayadhi L, Mohammed A. Impact of auditory integrative training on transforming growth factor-β 1 and its effect on behavioral and social emotions in children with autism spectrum disorder. Med Princ Pract. 2018;11461:23–9.
22. Ramaci T, Barattucci M, Ledda C, Rapisarda V. Social stigma during COVID-19 and its impact on HCWs outcomes. Sustainability. 2020;12(9):1–13.

Exploring the Role of the Rehabilitation Care Team in the Autistic Child

3

Baljeet Maini

3.1 Introduction

Team management is commonly followed for many chronic conditions in the healthcare sector. Autism is a complex lifelong condition and affected children have a variety of chronic symptoms and comorbidities. After diagnosis, all patients of autism should be assessed thoroughly by a variety of specialists. Also, to be most effective, the management interventions of autism patients should be carried out in a collaborated approach by a rehabilitation team of specialists from various disciplines.

A paediatrician is usually the primary physician of contact in most of the situations. But once diagnosed, the management and rehabilitation of a case of autism goes well beyond the domain of a paediatrician. Broadly, the important aspects of autism management (in accordance with latest Diagnostic and Statistical Manual [DSM]-5 criteria) are:

1. As much as possible, to facilitate and stimulate the normal development of cognitive function, language usage, and socialization processes.
2. To attenuate autism-related maladaptive behaviours like rigidity and stereotypy.
3. To reduce nonspecific behaviours of maladaptive nature such as hyperactivity, impulsivity, and irritability.
4. To reduce stress and burden for the family of the autism patient [1].

Many other associated (medical, mental, behavioural) issues, which are very commonly seen accompanying autism, also need to be diagnosed and addressed properly. It is now well established that autism cases need to be rehabilitated across

B. Maini (✉)
Department of Pediatrics, Maharishi Markandeshwar Institute of Medical Sciences and Research, Maharishi Markandeshwar (Deemed to be University), Mullana - Ambala, Haryana, India
e-mail: mainibaljeet@mmumullana.org

M. Goyal, K. Goyal (eds.), *Rehabilitation Approach in Autism*,
https://doi.org/10.1007/978-981-96-4162-8_3

a well-coordinated multidisciplinary approach, much beyond the scope of a primary caregiver office [2]. In a recent article from India, authors recommend that the assessment and rehabilitation of a suspected case should as far as possible be done by a team including a psychiatrist, psychologist, an occupational therapist, an audiologist, speech therapist, and, if available, a special educator for autism. In the western world, educational psychologists in school settings have been reported to be very helpful for the assessment of cognitive and curricular functions too, in bigger children with autism [3]. In a recent study from Romania, the authors concluded that a better understanding of comorbidities in ASD patients develops through interdisciplinary collaboration, thus leading to effective treatment programmes [4].

So, at present world-over the rehabilitation of a case of autism is now proven to be best undertaken by a team of experts. This team is not only a group of individuals working 'on their own' but in the best possible coordination among themselves for the best outcomes in a customized fashion.

3.2 Concept of Team Approach in the Rehabilitation of Autism

Autism patients have lots of needs in a varying manner. In today's world, a complex issue like autism is better manageable but only with a concerted effort by various kinds of specialists. It is a common understanding that it is not possible for any single person to be completely trained in all required skills and abilities. Within each area of healthcare, there are further sub-specializations, for instance, in medicine, practitioners do specialize in diverse medical specialities. Likewise, many other professionals, e.g. social workers and educationists, do specialize in their own sub-specialities. But together, in a multidisciplinary rehabilitation team, they represent specific areas of their special skills and capabilities with the aim that the ultimate user (the patient) benefits from an understanding that combines all fields and extends way beyond individual professional boundaries. Most of the experience in the multidisciplinary rehabilitation team for autism is based from western developed nations of America and Europe. This approach has been found to be very efficacious and is the recommended strategy as per the latest autism management guidelines [2]. Also the multidisciplinary management has been covered by various public healthcare funding programmes for autism patients. Many underdeveloped nations still do not have a well-established multidisciplinary team approach for autism and experience is still evolving.

3.3 Benefits of Team Approach in Autism Rehabilitation

3.3.1 For Patients

It has been realized very well now that autism patients benefit much more with a multidisciplinary team approach for their overall management of all medical and mental health.

(a) *Patient Access to Whole Care Team*

Patients of autism have a large variety of medical, psychological, and social needs to be taken care of. Specialists of a wide range of healthcare and social sectors can be accessed by the patient, when using a team approach.

(b) *Time-Saving and Streamlined Workflow*

Autism patients not only get the benefits of many specialists of various kinds, but also the added benefits of that being provided in a much less time- and money-consuming way. More and more autism patients also are now being managed with multidisciplinary approach for last few decades, and in some countries, now it is a legal requirement to treat the patients with autism using a multidisciplinary team approach. For instance, it is now a method in usage as per the National Institute of Clinical Excellence (NICE) guidelines 2013 [5, 6] for the management of autism, with a central idea that the 'whole is greater than the sum of its parts' [7].

(c) *Patient Can Set Goals for Himself*

In today's world, the aims and goals of patient treatment (especially for chronic conditions like autism) are set by the patients/their families (especially in cases of children and debilitated patients). A lot of factors account for the treatment approach for autism even in a team of health workers, but patients'/guardian's (in some cases) own participation has now been placed at a far important position in all multidisciplinary approaches and autism is no exception now.

(d) *Enhanced Team and Patients' Coordination*

In a multidisciplinary team approach, it is not necessary that all patients will be treated in a set sequence of diagnostic/therapeutic standpoints. An autism patient can have some unique issues and the same are taken into account in approach of the treating team. Issues can range from medical to social which affect the sequence of patient care. Simultaneously, the availability of specialists may be affected due to some reason and that can be coordinated for tailoring any given patient's treatment.

3.3.2 Benefits for Rehabilitation Team Members

The output of the multidisciplinary team is far greater than the simple addition of output given by each individual specialist. Though a paediatrician, a psychologist, and a speech therapist are at the core of the team for autism patients, but due to many medical and mental issues, a lot of other kinds of specialists are needed. Healthcare members benefit a lot in team approach because understanding each other's viewpoint becomes far easy as well as communicating the same to other team members. Beyond medical and psychology, the teams for autism consist of allied health sectors like occupational/physical therapy, speech therapist, behavioral analyst, clinical psychologist, social worker, special eductaion teacher, etc. as well as specialists from social care and educational fields.

3.3.3 Overall Enhanced Satisfaction

Overall, the multidisciplinary team approach brings in far greater benefits for patients as well as healthcare professionals. Not only is it medically more productive in output in the short term, patients with autism have been reported to have more satisfying lives in adolescent and adult age [3].

3.4 Role of Different Specialists in Autism Rehabilitation Team

The following discussion is about the individual roles and responsibilities of various kinds of specialities in a team for rehabilitation of cases of autism. In today's era, the parents are, and henceforth shall always be, at the core of the team rehabilitating any individual case of autism, in the entire network of care.

3.4.1 The Paediatrician

The paediatrician is usually the primary person for autism suspicion and screening. The paediatrician also guides and counsels the parents with appropriate initial referral to relevant specialists and regular monitoring to note changes in the child in a continuous manner [8]. He/she also has the responsibility of addressing other medical issues associated (or seen to be more common) with autistic children [2]. A paediatrician is the core solution provider for nutritional and vaccination issues besides other common medical issues of children. A paediatrician also acts as central coordinator of all healthcare needs of an autistic child, in most of autism rehabilitation teams.

3.4.2 The Psychologist

A child psychologist has the responsibility of many core issues related to autistic children. Each case of autism needs a proper and valid testing of cognitive functions. There are established standardized tools for determining the developmental levels of smaller children and intelligence quotient (IQ) in children more than 3 years of age. The psychologist selects an intelligence testing tool depending upon the age and linguistic ability of the child. Administering a valid cognitive test helps not only in ascribing symptoms to autism as part of the initial diagnosis, but also helps to establish other coexisting diagnoses, such as intellectual disability [9].

There are some valid tools which can be applied even to nonverbal children. Thus, it is evident that these important unique aspects need to be taken care of by a certified expert in this field, which is usually a child psychologist. He/she also assess the behavioural issues associated with these children and also provides advice and guidance to address them. He/she is the key person in teaching/training these children (and their parents) for continuous improvement in the condition and in achieving the goals of treatment (with the exception of co-existing medical conditions). In fact, it is the psychologist who is associated closest with the paediatrician/primary healthcare provider in managing an autistic child [10].

3.4.3 The Speech Pathologist/Therapist

Very typically, some differences are observed in the use of verbal and nonverbal communication (for purpose of social interaction) in any autistic child. There is a need of formal assessment of communication ability of the child by a speech therapist, at the time of initial diagnosis, and should always include (as far as possible) noting the levels of expressive and receptive linguistic skills along with use of conventional language [11].

3.5 Other Team Members in Autism Rehabilitation

It is very clear from the above discussion that individual roles played by a paediatrician, a psychologist, and a speech therapist are at the core of autism management. All three are involved in screening, diagnosing the case, and associated behavioural and speech/communication difficulties in affected children. Beyond this, they also formulate the teaching/training of families/parents of children in achieving the treatment goals of autism, along with a continuous monitoring of the same on a longitudinal basis, across the entire childhood and beyond into adulthood. But autistic children have discreet special needs in addition to the paediatrician, psychologist, and speech therapist. Because of these needs, many specialists from diverse fields are needed to be involved. Discussion on these specialities as part of autism rehabilitation follows.

3.5.1 The Occupational and Physical Therapist

Occupational therapy (OT) assists people with autism in various skills, e.g. cognitive, social, physical, and motor skills. Occupational therapy improves everyday abilities that enable the autism patient to become self-dependent and participative in a broader range of activities. OT programmes are directed towards playing skills, learning methods, and, most important of all, self-care. Strategies of OT also help to deal with sensory challenges which are of large concern in many autistic children. Occupational therapy begins by assessing the child's basal level of ability. Several areas are assessed, like how the person learns and plays and takes cares of self along with interaction with the surrounding environment. It detects the issues that limit the autistic person from participating in daily life activities. The therapist then decides the goals and strategies which enable the person to work upon key capabilities. Some common therapeutic targets include the ability of dressing, use of toilet, eating, etc. independently. Sensory integration (SI) therapy is used in isolation or as part of a bigger programme of OT for autistic children. The difficulties of neurologic processing of information and integration of sensory inputs are overcome by SI, thus making the child's social interaction better. 'Sensory' activities are useful as part of an overall programme to make the child calm, to bring and reinforce a desired change in behaviour, and also to help with transitions between various activities [12].

In bigger children, occupational therapists also help in improving the development of playing skills, modifying the design of classroom and teaching schedules to enhance the abilities of attention and organization, and even providing training of some prevocational skill. More elaborate discussion on occupation and physical therapy for autism follows in the next few chapters of this book.

3.5.2 The Paediatric Dentist

Children with autism commonly have unmet dental needs more commonly than the normal population. The most commonly reported barrier for dental care is the difficulty in cooperating with hygiene and professional care. A bad dental health and behaviour stereotypy can often create feeding problems leading to malnutrition and other health issues in children. The presence of painful lesions can potentially aggravate the behavioural abnormalities. For proper dental care a dentist should always be an integral part of the autism management team [13].

3.5.3 The Audiometrist

Children with language delay or inattention to language must have an evaluation of hearing as part of their initial evaluation [14]. Hearing deficit should always be considered in children with language delays, behaviour problem, and inattention. Appropriate amplification also maybe indicated in a few cases [15]. Undoubtedly, there should be a full attempt in ruling out any other cause of hearing deficit which can impede speech development and cognitive/social development of the child [16]. Thus, an audiometrist should always be included in the team of professionals caring for autism.

3.5.4 The Ophthalmologist

Visual function testing should always be considered in the initial evaluation of children, especially those who are visually inattentive or have stereotypical behaviours involving the eyes (e.g. poking of eyes), or children who are not making any eye contact. Decreased visual acuity may affect badly in the interactive gaze activity-based teaching techniques, e.g. joint attention. It may require additional accommodations in the educational setting [17].

3.5.5 Psychiatric Evaluation

Patients with autism are more frequently found to be having behaviour abnormalities than the normal population. Most of these are amenable to behaviour modification techniques embedded in the teaching training programmes formulated by the psychologist. But these are sometimes beyond the limits of non-pharmacological methods (often classifiable as major morbidities like ADHD, depression, etc). It is prudent thus to have a psychiatrist to be placed in the team approach of autism management. Some children definitely need psychopharmacological agents for correction of serious behaviour difficulties and mental conditions found in many of autistic children [18]. The correction of mood disorders and sleep abnormalities is also very vital for the rehabilitation of a child with autism. Individualized approach along with involvement of the psychiatrist with family members has been proven to be very effective in the rehabilitation of psychiatric abnormalities in autism [19].

3.5.6 Paediatric Neurologist

Neuro-assessment of an autistic child may be indicated in the presence of some clinical features like abnormal neurologic examination, seizures, and headache. Radiological studies like MRI are helpful in understanding the neurological lesions. *Electroencephalogram* in autism patients often require facilities like sleep-deprived electroencephalogram (EEG), usually indicated when there is history of spells or abnormal behaviours frankly suggestive of seizures. Sometimes there is a history of regression in language skills, e.g. in disorders like acquired epileptic aphasia in Landau–Kleffner (LK) syndrome [20, 21].

3.5.7 Special Educator for Autism Children

Education means gaining of knowledge and skills to help a child gain independence in day-to-day life as well as personal care; it means not only learning in academics but also skills for socialization, communication, and adaptive abilities and decreasing the interfering behaviours and, at the same time, generalization of abilities in multiple situations and circumstances. Children with ASD need a well-structured educational plan and approach with special methods and techniques of teaching. Interventions should not only be well planned but also should be individualized along with being intensive. A reliable evaluation of the child's strengths is needed, with a focussed and detailed description of intervention goals and procedures along with monitoring of effectiveness. The Early Start Denver Model (ESDM) [22] and Treatment and Education of Autistic and Related Communication-Handicapped Children (TEACCH) [23] have both been found to be effective. In the presence of special education service providers, clinicians can very often guide families to empirically supported practices and also help them evaluate the usefulness of the educational services that are being offered [2]. Special educators are the key members in the rehabilitation team of children with autism. Owing to special features of abnormal speech and communication abilities of autistic children, the special skill of education demands a different setting of teaching as well as a differently specially trained teacher. There is a 'no-one-size-fits-all' approach in teaching an autistic child. Special education teachers assess the needs of individual children as well as continuously monitors the progress of the child. Thus, he/she becomes a key member of the multidisciplinary team managing a child of school age. For example, the therapeutic approach by the psychologist, psychiatrist, and occupational therapist may be modified by the inputs provided by the special education teacher. In a recent study the authors found a deficiency in awareness about autism in special educators of south Karnataka, India [24, 25]. With the increasing prevalence rate of autism reported worldwide, there is a large need of special education teachers particularly in developing nations.

3.5.8 Role of Social Worker

It has been observed that many people in the autism spectrum perform very well in tasks that need systematic thinking and repetition [26, 27]. Qualitative research with the families of people on the autism spectrum has also highlighted how people on the spectrum enrich their lives, often providing a source of happiness and inspiration [28]. However, on the contrary, these strengths may remain unrecognized by society, which leads to decreased social opportunities, increasing the social isolation further [29, 30]. Bishop-Fitzpatrick et al. strongly advocate that social workers have a great scope of looking after autistic spectrum patients in long-term social rehabilitation [31]. They also recommend taking social rehabilitation to the 'grand challenge' level.

Social workers create a bridge between the healthcare rehabilitation teams and the society (including the patients' families). Notably, with changes in social work education, accompanied with initiatives for autistic children and adults, an approach is necessary to increase the workforce of social workers who are especially trained to work with autism children.

3.5.9 Parents and the Family of Children with Autism

Recently, there has been a strong emphasis on the role of family and parents especially in intervention programmes in rehabilitation teams. There is a paucity of literature regarding the efficacy of parent-mediated approaches for outcomes like language and communication, child initiation and adaptive behaviour, and parents' stress [32]. However, there is strong evidence indicating positive change in parent-child interaction with family-centred approaches (e.g. shared attention or parent-child synchrony). Active engagement of families/caregivers as a form of co-therapy is desirable but it requires appropriate supervision and monitoring. Parents not only help in setting goals and priorities for their child's treatment, but they also help to teach/reinforce new skills at home. Parent-mediated interventions have been reported to be cost-effective and increase the sense of empowerment on the part of caregivers [12, 33].

References

1. Shenoy MD, Indla V, Reddy H. Comprehensive management of autism: current evidence. Indian J Psychol Med. 2017;39:727–31.
2. Dalwai S, Ahmed S, Udani V, Mundkur N, Kamath SS, Nair MKC, et al. Consensus statement of the Indian academy of pediatrics on evaluation and management of autism spectrum disorder. Indian Pediatr. 2017;54(5):385–93.
3. Subramanyam AA, Mukherjee A, Dave M, Chavda K. Clinical practice guidelines for autism spectrum disorders. Indian J Psychiatry. 2019;61(Suppl 2):254–69.
4. Stanciu AC, Rad F, Mihailescu I, Mateescu L, Grozavescu R, Andrei E, et al. Multidisciplinary approach in children with autism spectrum disorder. Eur Psychiatry. 2021;64(Suppl 1):S86.
5. National Institute for Health Care Excellence (NICE). Autism in General Practice (Royal College of General Practitioners) [Internet]. 2013. Available from: http://www.nice.org.uk/usingguidance/education/AutismInGeneralPractice.jsp
6. National Institute for Health Care Excellence (NICE). Autism: Management of autism in children and young people [Internet]. 2013. Available from: www.nice.org.uk/nicemedia/live/14257/64946/64946.pdf
7. Bond C, Eldridge S. SAPC hot topic—a challenge of identity for Academic Primary Care: keeping the whole greater than the sum of its parts. Prim Health Care Res Dev. 2012;13(4):403–5.
8. Committee on Children with Disabilities. American Academy of Pediatrics: the pediatrician's role in the diagnosis and management of autistic spectrum disorder in children. Pediatrics. 2001;107(5):1221–6.
9. Bryson SE, Bradley EA, Thompson A, Wainwright A. Prevalence of autism among adolescents with intellectual disabilities. Can J Psychiatr. 2008;53(7):449–59.

10. Reszka SS, Boyd BA, McBee M. Brief report: concurrent validity of autism symptom severity measures. J Autism Dev Disord. 2014;44(2):466–70.
11. O'Hare A, Bremner L. Management of developmental speech and language disorders: part 1. Arch Dis Child. 2016;101(3):272–7.
12. Myers SM, Johnson CP, American Academy of Pediatrics Council on Children with Disabilities. Management of children with autism spectrum disorders. Pediatrics. 2007;120(5):1162–82.
13. Du RY, Yiu CKY, King NM. Oral health behaviours of preschool children with autism spectrum disorders and their barriers to dental care. J Autism Dev Disord. 2019;49(2):453–9.
14. Bennetto L, Keith JM, Allen PD, Luebke AE. Children with autism spectrum disorder have reduced otoacoustic emissions at the 1 kHz mid-frequency region. Autism Res. 2017;10(2):337–45.
15. Dawes P, Bishop D. Auditory processing disorder in relation to developmental disorders of language, communication and attention: a review and critique. Int J Lang Commun Disord. 2009;44(4):440–65.
16. Hyman SL, Levy SE, Myers SM, Council on Children with Disabilities, Section on Developmental and Behavioral Pediatrics. Identification, evaluation, and management of children with autism spectrum disorder. Pediatrics. 2020;145(1):e20193447.
17. Anketell PM, Saunders KJ, Gallagher SM, Bailey C, Little JA. Vision in children with autism spectrum disorder: what should clinicians expect? J Autism Dev Disord. 2015;45(9):3041–7.
18. Buck TR, Viskochil J, Farley M, Coon H, McMahon WM, Morgan J, et al. Psychiatric comorbidity and medication use in adults with autism spectrum disorder. J Autism Dev Disord. 2014;44(12):3063–71.
19. Kreslins A, Robertson AE, Melville C. The effectiveness of psychosocial interventions for anxiety in children and adolescents with autism spectrum disorder: a systematic review and meta-analysis. Child Adolesc Psychiatry Ment Health. 2015;9:22.
20. El Achkar CM, Spence SJ. Clinical characteristics of children and young adults with co-occurring autism spectrum disorder and epilepsy. Epilepsy Behav. 2015;47:183–90.
21. National Research Council. Educating children with autism. Washington, DC: National Academy of Sciences Press; 2001.
22. Dawson G, Rogers S, Munson J, Smith M, Winter J, Greenson J, et al. Randomized, controlled trial of an intervention for toddlers with autism: the Early Start Denver Model. Pediatrics. 2010;125(1):e17–23.
23. Mesibov GB, Shea V, Schopler E. The TEACCH approach to autism spectrum disorders. New York: Kluwer Academic/Plenum; 2005.
24. Oono IP, Honey EJ, McConachie H. Parent-mediated early intervention for young children with autism spectrum disorders (ASD). Cochrane Database Syst Rev. 2013;4:CD009774.
25. Tiwari S, John J. Special educators' knowledge and training on autism in Karnataka: a cross-sectional study. Indian J Psychiatry. 2017;59(3):359–65.
26. Baron-Cohen S, Ring HA, Wheelwright S, Bullmore ET, Brammer MJ, Simmons A, et al. Social intelligence in the normal and autistic brain: an fMRI study. Eur J Neurosci. 1999;11(6):1891–8.
27. Ozonoff S, Cathcart K. Effectiveness of a home program intervention for young children with autism. J Autism Dev Disord. 1998;28(1):25–32.
28. Timmons L, Ekas NV, Johnson P. Thankful thinking: a thematic analysis of gratitude letters by mothers of children with autism spectrum disorder. Res Autism Spectr Disord. 2017;34:19–27.
29. O'Malley-Keighran MP, Coleman M. 'I am not a tragedy. I am full of hope': communication impairment narratives in newspapers. Int J Lang Commun Disord. 2014;49(2):174–88.
30. Wolbring G, Mosig K. Autism in the news: content analysis of autism coverage in Canadian newspapers. In: Baker DL, editor. Disability and US politics: participation, policy, and controversy. Santa Barbara: ABC-CLIO; 2017. p. 63–94.
31. Bishop-Fitzpatrick L, Dababnah S, Baker-Ericzén MJ, Smith MJ, Magaña SM. Autism spectrum disorder and the science of social work: a grand challenge for social work research. Soc Work Ment Health. 2019;17(1):73–92.

32. Zwaigenbaum L, Bauman ML, Choueiri R, Kasari C, Carter A, Granpeesheh D, et al. Early intervention for children with autism spectrum disorder under 3 years of age: recommendations for practice and research. Pediatrics. 2015;136(Suppl 1):S60–81.
33. Volkmar F, Siegel M, Woodbury-Smith M, et al. Academy of Child and Adolescent Psychiatry (AACAP) Committee on Quality Issues (CQI). Practice parameter for the assessment and treatment of children and adolescents with autism spectrum disorder. J Am Acad Child Adolesc Psychiatry. 2014;53(2):237–57.

Introduction to Evidence-Based Treatment Options in Autistic Children

4

Hina Vaish

4.1 Introduction

In the present scenario autism is the most extensively debated condition. The augmented prevalence of autism has made it more widely known in society [1].

Autism symptoms often appear in childhood, but not all may develop until society's expectations of the individuals' abilities are high. Individuals with autism have different needs and abilities, and these might change over time. While some autistic individuals can live independently, others have significant difficulties and need care and support for the rest of their lives [2].

The necessity for effective therapeutic facilities has grown along with the increased prevalence of autism [3], and intervention science is increasingly supplying information about strategies that improve results. Early management strategies are required to positively impact individuals with autism [4]. There are various sorts of treatment options available, although some management strategies involve more than one approach. Numerous therapies are available for individuals with autism in learning new abilities and overcoming a range of developmental challenges. Existing management strategies for autism aim to reduce symptoms that impact everyday activities and quality of life (QoL) [5]. Because information regarding the characteristics of autistic individuals, their abilities, and management approaches is constantly evolving, it is necessary to keep up to date on evidence-based practices (EBPs) [6].

Integrated treatment approaches should be used in individuals with autism. The treatment of individuals with autism should be tailored as the treatment objectives will vary according to the patient's cognition level and functional abilities. The purpose of treatment strategies adopted must be to maximize the QoL constantly for the individuals with autism [7].

H. Vaish (✉)
School of Health Sciences, Chhatrapati Shahu Ji Maharaj University, Kanpur, India
e-mail: hinavaish@csjmu.ac.in

M. Goyal, K. Goyal (eds.), *Rehabilitation Approach in Autism*,
https://doi.org/10.1007/978-981-96-4162-8_4

In the present chapter there is an overview of current evidence-based treatment options for autism.

4.2 Evidence-Based Treatment Options

There exist different kinds of evidence-based management strategies available for autism. It has been recommended that as the needs and goals of treatment in individuals of autism are specific and varied, hence treatment strategies adopted should be individualized and a combination of treatment strategies adopted will lead to better and positive health outcomes. These treatments are presented in Fig. 4.1.

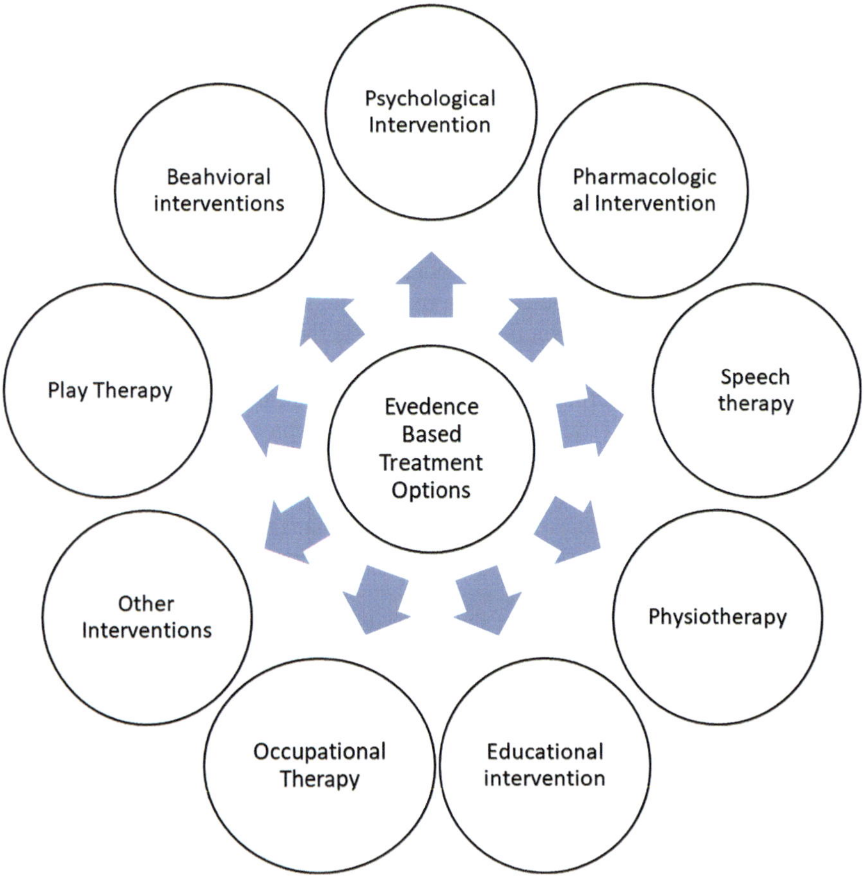

Fig. 4.1 Evidence-based treatment options for autism. [8–56]

4.2.1 Pharmacological Intervention

It includes different interventions by the drugs. The primary goal of pharmacological treatment for autism is to lessen the common symptoms such as irritability, aggression, repetitive behaviour, cognitive disorder, compulsions, impulsivity, hyperactivity, and inattention [8].

A variety of pharmaceutical treatments have been assessed in patients with ASD and through clinical trials [8]. Methylphenidate may be useful in reducing hyperactivity and inattention; risperidone and aripiprazole have demonstrated efficacy in treating irritability and aggression; melatonin is frequently used in treating sleep issues in individuals with autism [9]. However, it is recommended that pharmacological therapies ought to be provided in accordance with each patient's specific needs, taking into account both possible risks and advantages [8].

4.2.2 Behavioural Interventions

Behavioural interventions are proposed to reduce anxiety and aggression while improving language, cognitive function, adaptive behaviour, and social skills [10] (Fig. 4.2).

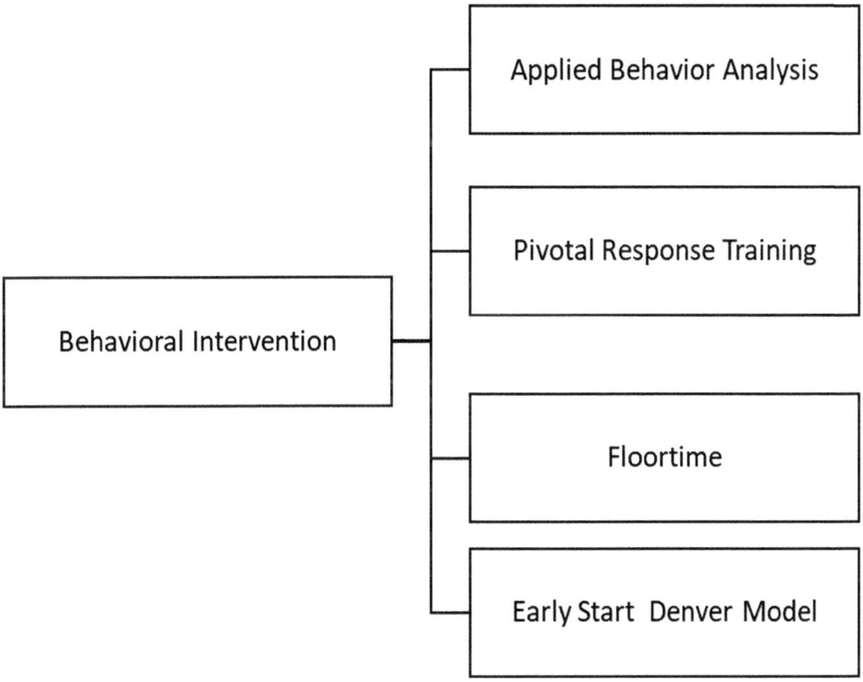

Fig. 4.2 Types of behavioural intervention used in autism. [11–19]

Applied Behaviour Analysis (ABA): There is vast evidence supporting the use of early intensive behavioural interventions (EIBIs) in individuals with autism [11, 12]. Based on behaviour analysis, ABA, also known as Discrete Trial Training, is among the most discussed types of behavioural intervention therapy. Substantial repetition in addition to reinforcement is used to teach complex behaviours in a systematic manner over time by breaking them down into smaller, discrete skills [13].

Pivotal Response Training (PRT): Based on fundamentals of ABA models, it helps individuals to learn new things and develop new habits that are crucial for the development of new skills and behaviours. These areas include an individual's motivation and cue-response behaviour. The use of tangible reinforcers in DTT is replaced with an emphasis on natural reinforcers [14].

Floortime: The Developmental, Individual Difference, Relationship-Based (DIR) model serves as the foundation for floortime, an early intervention framework that emphasizes a child's comprehensive social-emotional development over the acquisition of discrete or isolated skills. Most of the interventions is based on play activities and concentrates on actions that suit the interests of the child and the stage of development [16, 17].

Early Start Denver Model (ESDM): Over time, it has been observed that ESDM lessens the severity of autism symptoms by combining aspects of structured ABA with less formal relationship-based approaches to intervention [18, 19].

It has been proposed that it works better to combine behavioural interventions with medication for better outcomes [10].

4.2.3 Physiotherapy

Children with autism have been found to have motor delays and deficits. These issues may worsen with time, relapse into groups of chronic disorder, or spread more widely as they get older [20, 21]. Exercise is generally thought to lower the risk of wide-ranging health issues in individuals of autism in a similar manner as in general population. Exercise has several benefits in individuals with autism as shown in Table 4.1. Additionally, researchers have indicated that exercise has effects in improving balance, postural control, and motor skills [22–24].

Physiotherapy aids autistic children in maintaining and improving the deficits and potential future motor deficits. Authors report that individuals with autism

Table 4.1 Benefits of physical therapy in individuals with autism [22–34]

Improves motor skills
Improves balance
Improves coordination
Improves postural control
Reduces sensory issues
Improves social interaction
Improves behavioural issues
Enhances functional performance
Improves overall health

experience improvements in their physical environments and abilities to communicate, which helps to improve their cognitive and core disabilities [25]. Children who receive physical therapy not only benefit from exercise but also develop independence [26].

Physical exercise enhances physical condition, cognitive and sensorial capacities, motor performance, and coordination in this population, including arm function, balance, movement planning, and gait [23]. Authors from another study investigated the impact of short-term training effect of rehabilitation programmes on postural control parameters in patients with autism and reported that results show mixed effects of training rehabilitation and test-retest, and visual inputs improve postural sway [27].

Physical exercises in the form of strengthening, aerobic exercises, balance training, and manual techniques are fundamental in improving symptoms of autism [23, 28–30]. The techniques like climbing & supporting bars, using baskets & medicinal balls, and elastic cuffs are used to facilitate strengthening in the upper & lower limbs in an autistic child [31]. Encouraging the child to take objects can help maintain their sensory system. Skateboarding can help develop new motor skills and enhance mental and executive functions, while exercise games can decrease stereotyped behaviours in individuals with autism [32].

Aquatic therapy has substantial benefits because of the water properties of hydrostatic pressure and buoyancy [33]. It appears to be predominantly advantageous for children with autism as in them strong sensory stimuli is needed [34].

The application of physiotherapy as a treatment requires careful consideration of individual needs, preferences, and talents [35].

4.2.4 Occupational Therapy

Delays and persistent deficits in fine moor skills and gross motor skills is a common presentation in people with autism [36, 37]. The use of occupational therapy is beneficial in improving the gross and fine motor deficits, as well as improve early cognitive development in people with autism [38]. The aim of occupational therapy in autism is to promote work-related engagement and social participation in day-to-day environments, enabling them to perform tasks required for their roles [39].

Some people with autism exhibit differences in their sensory processing that needs a particular intervention. However, there is little empirical evidence in favour of sensory integration therapy [40]. Occupational therapists frequently use sensory integration techniques for people with autism [41]. The benefits of sensory integration treatment include improved motor abilities, socializing, focus, behaviour control, reading comprehension, game participation, and the ability to reach personal goals [42].

4.2.5 Educational Interventions

Special Education Services Children with autism need a lot of help during their schooling. An individualized education programme (IEP) is frequently a crucial component in arranging and overseeing the proper academic development of a student. As early as 3 years old, a student may be placed on an IEP, and many continue in formal education programmes until they are 22 years old.

Treatment and Education of Autistic and Related Communications Handicapped Children (TEACCH) This model is a collection of instructional and involvement techniques that are based on the unique learning necessities for individuals with autism, which may include challenges with social communication, diminished executive functioning, and a strength in visual processing. Recent studies report some benefit in the development of verbal, motor, perceptual, and cognitive skills as well as enhanced behaviour and social skills [43].

4.2.6 Psychosocial Interventions

Parent Management Training Researches across the world show that parent management training is beneficial in lowering stress levels in parents and enhancing the regulation of challenging behaviours in individuals with autism, including aggression, non-compliance, and self-harm [44, 45]. Furthermore, it has been discovered that parent education plus medication reduces severe maladaptive behaviours more than medication alone and, in certain situations, even permits a reduction in medication dosage [46].

Cognitive Behavioural Therapy (CBT) Of all the psychotherapies, CBT is one of the most widely used and thoroughly researched; it is a fusion of behavioural therapy and cognitive therapy, two different therapeutic modalities [47]. Studies indicate that CBT helps people with ASD experience fewer concurrent psychiatric symptoms, like anxiety [48].

Family Therapy Psychological distress, marital issues, and family conflict are frequently experienced by parents and children with autism [49, 50]. A family system's general well-being may be supported by family therapy, which includes couple counselling.

4.2.7 Speech Therapy

Autistic individuals face language and communication challenges through life such as delay in the development of language, poor articulation, and lack of social communication skills. To address the numerous speech- and language-related issues that autistic people face, speech therapy is an important intervention. It has been reported that speech therapy leads to improvement in achieving more spontaneous speech, enhanced classroom learning skills, social communication, and overall conversational competence [51, 52].

Speech language pathologists help individuals with autism to communicate effectively and interact successfully by enhancing oral motor skills, sensory concerns, and safe swallowing, thereby significantly influencing communication skills, social relationships, and language development [34].

4.2.8 Nutritional Intervention

Gastrointestinal symptoms are common presentation in the autistic population [53]. Gluten-free diets and casein-free diets are commonly consumed, though evidence is limited for their use [54]. Nutritional deficiencies or excess is a commonly a risk in patients with autism. Getting enough calcium is a specific concern but there is not much evidence to support daily consumption of vitamins and supplements [53]. Weight gain is commonly present among individuals with autism, especially in those who consume atypical antipsychotics which can influence their quality of life [55].

4.2.9 Play Therapy

Play therapy is commonly used to treat a variety of illnesses since years. Play therapy assists kids with autism in finding play activities that they enjoy, expressing themselves in ways that feel comfortable, and transitioning from undesirable to more constructive forms of self-expression [56].

Play therapy is an effective management strategy for autism, as it fosters trust, understanding of different interaction styles, and self-regulation [57]. It provides a safe environment for children to learn appropriate behaviours and allows them freedom of choice, promoting independence and developmental stimulation [31].

Play therapy is of different types as shown in Fig. 4.3.

Play therapy's likely impact on patients of autism and their families is the need to be evaluated through more well-designed controlled studies, despite weak evidence [56].

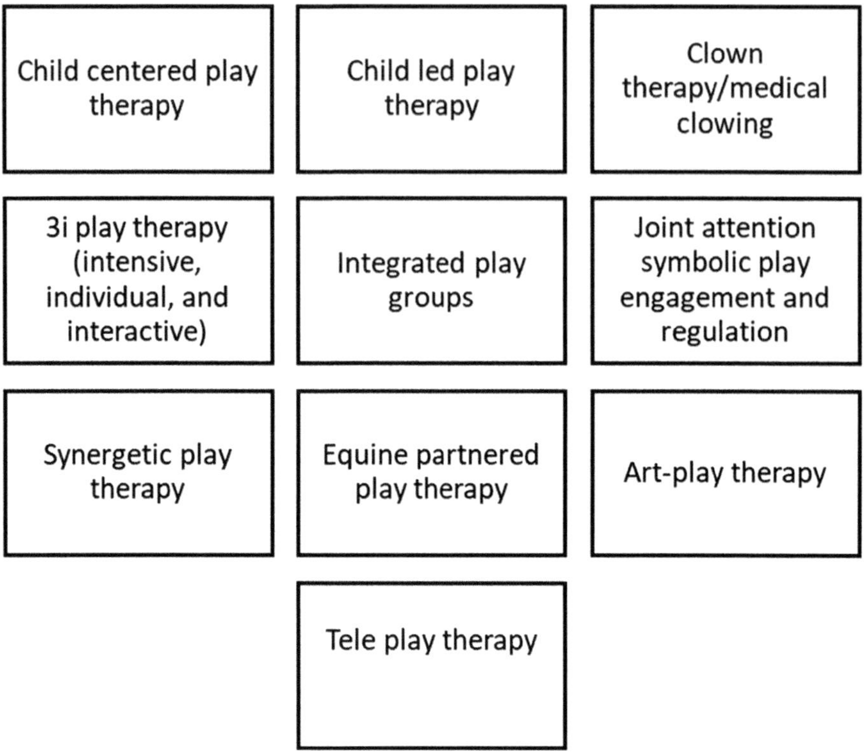

Fig. 4.3 Types of play therapy. [56]

4.2.10 Other Interventions

Hyperbaric Oxygen Therapy HBOT increases the levels of oxygen in the body that reduces inflammation and oxidative stress. HBOT may improve inflammation and oxidative stress, thus leading to improvement in symptoms of autism [58]. There is not much evidence of effectiveness of HBOP which is presented in few case series and few controlled trials with limited randomized, controlled trials having a positive effect [56]. The results of the other randomized, controlled trials are not in favour of the previous findings [12]. Presently, HBOT therapy for autism is not recommended.

Electroconvulsive Therapy for Catatonia (ECT) ECT is effective as a management strategy in treating depression and catatonia. Only few case reports and case series are there as most of the supporting evidence of ECT in treating autism-related catatonia; thus, more extensive randomized controlled trials are needed for generating the evidence [59].

Assistive Devices/Technology People diagnosed with autism frequently experience language impairments. Communication breakdowns can lead to or worsen behavioural issues like aggressiveness and chronic irritability. Augmentative communication protocols are developed to support language acquisition and provide a way of communication for people who have limited or no verbal language skills.

The picture exchange communication system is a popular tool; it is reported as a useful strategy in observational studies and randomized trials, especially in early childhood [60].

The advancement of software and devices that can generate speech has accelerated due to technological advancements in touchscreen handheld devices [61]. There is growing evidence that AAC devices are beneficial for children who are not very talkative [62].

4.3 Conclusion

Throughout the lifespan, autism requires a multidisciplinary approach to treatment owing to the complex neurodevelopmental disorder. The most significant factors in treating autism are those related to education, behaviour, and rehabilitation. Comprehensive treatment should be provided in specialized facilities. There are numerous types of evidence-based treatment options available presently to maximize autistic individuals' functional independence. Behavioural interventions are used in early childhood and has the strong evidence to treat the core social communication impairment in autism. Physiotherapy with its hands-on approach seems promising for augmenting QoL and positive and health outcomes in individuals with autism and have the potential to facilitate participation of autistic children in physical activity.

References

1. Lord C, Brugha TS, Charman T, et al. Autism spectrum disorder. Nat Rev Dis Primers. 2020;6(1):5. https://doi.org/10.1038/s41572-019-0138-4. Published 2020 Jan 16.
2. World Health Organization. 2023. Accessed from: Autism (who.int).
3. Maenner MJ, Shaw KA, Baio J, et al. Prevalence of autism spectrum disorder among children aged 8 years—autism and developmental disabilities monitoring network, 11 sites, United States, 2016 [published correction appears in MMWR Morb Mortal Wkly Rep. 2020 Apr 24;69(16):503. doi: 10.15585/mmwr.mm6916a4]. MMWR Surveill Summ. 2020;69(4):1–12. https://doi.org/10.15585/mmwr.ss6904a1. Published 2020 Mar 27.
4. Lai MC, Anagnostou E, Wiznitzer M, Allison C, Baron-Cohen S. Evidence-based support for autistic people across the lifespan: maximising potential, minimising barriers, and optimising the person-environment fit. Lancet Neurol. 2020;19(5):434–51. https://doi.org/10.1016/S1474-4422(20)30034-X.
5. Hyman SL, Levy SE, Myers SM, Council on Children with Disabilities, Section on Developmental and Behavioral Pediatrics. Identification, evaluation, and management of

children with autism spectrum disorder. Pediatrics. 2020;145(1):e20193447. https://doi.org/10.1542/peds.2019-3447.
6. Hume K, Steinbrenner JR, Odom SL, et al. Evidence-based practices for children, youth, and young adults with autism: third generation review [published correction appears in J Autism Dev Disord. 2023 Jan;53(1):514. doi: 10.1007/s10803-022-05438-w]. J Autism Dev Disord. 2021;51(11):4013–32. https://doi.org/10.1007/s10803-020-04844-2.
7. Politte LC, Howe Y, Nowinski L, et al. Evidence-based treatments for autism spectrum disorder. Curr Treat Options Psychiatry. 2015;2:38–56. https://doi.org/10.1007/s40501-015-0031-z.
8. LeClerc S, Easley D. Pharmacological therapies for autism spectrum disorder: a review. P T. 2015;40(6):389–97. Available from: https://pubmed.ncbi.nlm.nih.gov/26045648
9. Davico C, Secci I, Vendrametto V, et al. Pharmacological treatments in autism spectrum disorder: a narrative review. PSY [Internet]. 2023;29(1/2). [cited 2024 Mar. 24]. Available from: https://www.jpsychopathol.it/article/view/251
10. Dawson G, Burner K. Behavioral interventions in children and adolescents with autism spectrum disorder: a review of recent findings. Curr Opin Pediatr. 2011;23(6):616–20. https://doi.org/10.1097/MOP.0b013e32834cf082.
11. Lovaas OI, Smith T. Evidence-based psychotherapies for children and adolescents. In: Kazdin AE, Weisz JR, editors. The Guilford Press; 2003. Available from: https://psycnet.apa.org/record/2010-09488-000
12. Remington B, Hastings RP, Kovshoff H, et al. Early intensive behavioral intervention: outcomes for children with autism and their parents after two years. Am J Ment Retard. 2007;112(6):418–38. https://doi.org/10.1352/0895-8017(2007)112[418:EIBIOF]2.0.CO;2.
13. Granpeesheh D, Tarbox J, Dixon DR. Applied behavior analytic interventions for children with autism: a description and review of treatment research. Ann Clin Psychiatry. 2009;21(3):162–73. Available from: PMID: 19758537.
14. Koegel RL, Koegel LK, Vernon TW, et al. Empirically supported pivotal response treatment for children with autism spectrum disorders. In: Evidence-based psychotherapies for children and adolescents. New York; 2010.
15. Koegel LK, Ashbaugh K, Koegel RL. Pivotal response treatment. In: Early intervention for young children with autism spectrum disorder. 2016. p. 85–112. https://psycnet.apa.org/doi/10.1007/978-3-319-30925-5_4
16. Greenspan SI, Wieder S. Engaging autism: using the floortime approach to help children relate, communicate, and think. Da Capo Lifelong Books; 2006. Available from: https://api.semanticscholar.org/CorpusID:141515019
17. Greenspan SI, Wieder S. The developmental individual-difference, relationship-based (DIR/Floortime) model approach to autism spectrum disorders. 2007. Available from: https://psycnet.apa.org/record/2007-14932-010
18. Rogers SJ, Dawson G. Early Start Denver Model for young children with autism: promoting language, learning, and engagement. Guilford Publications; 2020.
19. Dawson G, Rogers S, Munson J, et al. Randomized, controlled trial of an intervention for toddlers with autism: the Early Start Denver Model. Pediatrics. 2010;125(1):e17–23. https://doi.org/10.1542/peds.2009-0958.
20. Ohta M, Nagai Y, Hara H, et al. Parental perception of behavioral symptoms in Japanese autistic children. J Autism Dev Disord. 1987;17(4):549–63. https://doi.org/10.1007/BF01486970.
21. Seynhaeve I, Nader-Grosbois N, Dionne C. Functional abilities and neuropsychological dysfunctions in young children with autism and with intellectual disabilities. Alternatives. 2008;2(3):230–52. https://doi.org/10.1016/j.alter.2008.05.001.
22. Hariri R, Nakhostin-Ansari A, Mohammadi F, et al. An overview of the available intervention strategies for postural balance control in individuals with autism spectrum disorder. Autism Res Treat. 2022;2022:3639352. https://doi.org/10.1155/2022/3639352. Published 2022 Nov 21.
23. Roşca AM, Rusu L, Marin MI, et al. Physical activity design for balance rehabilitation in children with autism spectrum disorder. Children (Basel). 2022;9(8):1152. https://doi.org/10.3390/children9081152. Published 2022 Jul 30.

24. Castaño PRL, Suárez DPM, González ER, et al. Effects of physical exercise on gross motor skills in children with autism spectrum disorder. J Autism Dev Disord. 2024;54(8):2816–25. https://doi.org/10.1007/s10803-023-06031-5.
25. Geslak DS. Challenging autism with exercise: an opportunity worth stretching for. ACSMs Health Fit J. 2017;21(2):19–24. https://doi.org/10.1249/FIT.0000000000000280.
26. Pan CY, Chu CH, Tsai CL, et al. The impacts of physical activity intervention on physical and cognitive outcomes in children with autism spectrum disorder. Autism. 2017;21(2):190–202. https://doi.org/10.1177/1362361316633562.
27. Caldani S, Atzori P, Peyre H, et al. Short rehabilitation training program may improve postural control in children with autism spectrum disorders: preliminary evidences. Sci Rep. 2020;10(1):7917. https://doi.org/10.1038/s41598-020-64922-4. Published 2020 May 13.
28. Coffey C, Carey M, Kinsella S, et al. Exercise programming for children with autism spectrum disorder: recommendations for strength and conditioning specialists. Strength Condition J. 2021;43(2):64–74. https://doi.org/10.1519/SSC.0000000000000599.
29. Oriel KN, George CL, Peckus R, et al. The effects of aerobic exercise on academic engagement in young children with autism spectrum disorder. Pediatr Phys Ther. 2011;23(2):187–93. https://doi.org/10.1097/PEP.0b013e318218f149.
30. Jungade S. Manual physical therapy as a novel treatment modality for autism spectrum disorder—a pilot study. J Complement Integr Med. 2020;17(2):/j/jcim.2020.17.issue-2/jcim-2018-0243/jcim-2018-0243.xml. https://doi.org/10.1515/jcim-2018-0243. Published 2020 Mar 31.
31. Donnellan AM, Hill DA, Leary MR. Rethinking autism: implications of sensory and movement differences for understanding and support. Front Integr Neurosci. 2013;6:124. https://doi.org/10.3389/fnint.2012.00124. Published 2013 Jan 28.
32. Watling R, Tomchek S, LaVesser P. The scope of occupational therapy services for individuals with autism spectrum disorders across the lifespan. AJOT Am J Occup Ther. 2005;59(6):680–4. https://doi.org/10.5014/ajot.59.6.680.
33. Becker BE. Aquatic therapy: scientific foundations and clinical rehabilitation applications. PM R. 2009;1(9):859–72. https://doi.org/10.1016/j.pmrj.2009.05.017.
34. Güeita-Rodríguez J, Ogonowska-Slodownik A, Morgulec-Adamowicz N, et al. Effects of aquatic therapy for children with autism spectrum disorder on social competence and quality of life: a mixed methods study. Int J Environ Res Public Health. 2021;18(6):3126. https://doi.org/10.3390/ijerph18063126. Published 2021 Mar 18.
35. Kalra R, Gupta M, Sharma P. Recent advancement in interventions for autism spectrum disorder: a review. J Neurorestoratol. 2023;21:100068. https://doi.org/10.1016/j.jnrt.2023.
36. Provost B, Lopez BR, Heimerl S. A comparison of motor delays in young children: autism spectrum disorder, developmental delay, and developmental concerns. J Autism Dev Disord. 2007;37(2):321–8. https://doi.org/10.1007/s10803-006-0170-6.
37. Liu T. Motor milestone development in young children with autism spectrum disorders: an exploratory study. Educ Psychol Pract. 2012;28(3):315–26. Available from: https://api.semanticscholar.org/CorpusID:144953016
38. Frolek Clark GJ, Schlabach TL. Systematic review of occupational therapy interventions to improve cognitive development in children ages birth-5 years. Am J Occup Ther. 2013;67(4):425–30. https://doi.org/10.5014/ajot.2013.006163.
39. Kashefimehr B, Kayihan H, Huri M. The effect of sensory integration therapy on occupational performance in children with autism. OTJR (Thorofare N J). 2018;38(2):75–83. https://doi.org/10.1177/1539449217743456.
40. Lang R, O'Reilly M, Healy O, et al. Sensory integration therapy for autism spectrum disorders: a systematic review. 2012. In: Database of Abstracts of Reviews of Effects (DARE): Quality-assessed Reviews [Internet]. York: Centre for Reviews and Dissemination (UK); 1995. Available from: https://www.ncbi.nlm.nih.gov/books/NBK100214/
41. Schaaf RC, Benevides TW, Kelly D, et al. Occupational therapy and sensory integration for children with autism: a feasibility, safety, acceptability and fidelity study. Autism. 2012;16(3):321–7. https://doi.org/10.1177/1362361311435157.

42. Schaaf RC, Hunt J, Benevides T. Occupational therapy using sensory integration to improve participation of a child with autism: a case report. Am J Occup Ther. 2012;66(5):547–55. https://doi.org/10.5014/ajot.2012.004473.
43. Virues-Ortega J, Julio FM, Pastor-Barriuso R. The TEACCH program for children and adults with autism: a meta-analysis of intervention studies. Clin Psychol Rev. 2013;33(8):940–53. https://doi.org/10.1016/j.cpr.2013.07.005.
44. Singh NN, Lancioni GE, Winton AS, et al. Mindful parenting decreases aggression, noncompliance, and self-injury in children with autism. J Emot Behav Disord. 2006;14(3):169–77. https://doi.org/10.1177/10634266060140030401.
45. Singh NN, Lancioni GE, Winton AS, et al. Training in mindful caregiving transfers to parent–child interactions. J Child Family Stud. 2010;19:167–74. Available from https://api.semanticscholar.org/CorpusID:144947825
46. Aman MG, McDougle CJ, Scahill L, et al. Medication and parent training in children with pervasive developmental disorders and serious behavior problems: results from a randomized clinical trial [published correction appears in J Am Acad Child Adolesc Psychiatry. 2010 Jul;49(7):727]. J Am Acad Child Adolesc Psychiatry. 2009;48(12):1143–54. https://doi.org/10.1097/CHI.0b013e3181bfd669.
47. Chand SP, Kuckel DP, Huecker MR. Cognitive behavior therapy. [Updated 2023 May 23]. In: StatPearls [Internet]. Treasure Island: StatPearls Publishing; 2024. Available from: https://www.ncbi.nlm.nih.gov/books/NBK470241/
48. Storch EA, Arnold EB, Lewin AB, et al. The effect of cognitive-behavioral therapy versus treatment as usual for anxiety in children with autism spectrum disorders: a randomized, controlled trial. J Am Acad Child Adolesc Psychiatry. 2013;52(2):132–42.e2. https://doi.org/10.1016/j.jaac.2012.11.007.
49. Hayes SA, Watson SL. The impact of parenting stress: a meta-analysis of studies comparing the experience of parenting stress in parents of children with and without autism spectrum disorder. J Autism Dev Disord. 2013;43(3):629–42. https://doi.org/10.1007/s10803-012-1604-y.
50. Lee GK, Lopata C, Volker MA, et al. Health-related quality of life of parents of children with high-functioning autism spectrum disorders. Focus Autism Other Dev Disabilities. 2009;24(4):227–39. https://doi.org/10.1177/1088357609347371.
51. Ingersoll B, Dvortcsak A, Whalen C, et al. The effects of a developmental, social—pragmatic language intervention on rate of expressive language production in young children with autistic spectrum disorders. Focus Autism Other Dev Disabilities. 2005;20(4):213–22. Available from: https://psycnet.apa.org/record/2005-16255-003
52. Adams C, Lockton E, Freed J, et al. The Social Communication Intervention Project: a randomized controlled trial of the effectiveness of speech and language therapy for school-age children who have pragmatic and social communication problems with or without autism spectrum disorder. Int J Lang Commun Disord. 2012;47(3):233–44. https://doi.org/10.1111/j.1460-6984.2011.00146.x.
53. Coury DL, Ashwood P, Fasano A, et al. Gastrointestinal conditions in children with autism spectrum disorder: developing a research agenda. Pediatrics. 2012;130(Suppl 2):S160–8. https://doi.org/10.1542/peds.2012-0900N.
54. Marí-Bauset S, Zazpe I, Mari-Sanchis A, et al. Evidence of the gluten-free and casein-free diet in autism spectrum disorders: a systematic review. J Child Neurol. 2014;29(12):1718–27. https://doi.org/10.1177/0883073814531330.
55. Curtin C, Jojic M, Bandini LG. Obesity in children with autism spectrum disorder. Harv Rev Psychiatry. 2014;22(2):93–103. https://doi.org/10.1097/HRP.0000000000000031.
56. Elbeltagi R, Al-Beltagi M, Saeed NK, et al. Play therapy in children with autism: its role, implications, and limitations. World. J Clin Pediatr. 2023;12(1):1–22. https://doi.org/10.5409/wjcp.v12.i1.1. Published 2023 Jan 9
57. Kennedy-Behr A, Rodger S, Mickan S. Aggressive interactions during free-play at preschool of children with and without developmental coordination disorder. Res Dev Disabil. 2013;34(9):2831–7. https://doi.org/10.1016/j.ridd.2013.05.033.

58. Bent S, Bertoglio K, Ashwood P, et al. Brief report: hyperbaric oxygen therapy (HBOT) in children with autism spectrum disorder: a clinical trial. J Autism Dev Disord. 2012;42(6):1127. Caldani S, Atzori P, Peyre H, Delorme R, Bucci MP. Short rehabilitation training program may improve postural control in children with autism spectrum disorders: preliminary evidences. Sci Rep. 2020;10(1):7917. https://doi.org/10.1038/s41598-020-64922-4-1132. https://doi.org/10.1007/s10803-011-1337-3. Published 2020 May 13.
59. DeJong H, Bunton P, Hare DJ. A systematic review of interventions used to treat catatonic symptoms in people with autistic spectrum disorders. J Autism Dev Disord. 2014;44(9):2127–36. https://doi.org/10.1007/s10803-014-2085-y.
60. Flippin M, Reszka S, Watson LR. Effectiveness of the Picture Exchange Communication System (PECS) on communication and speech for children with autism spectrum disorders: a meta-analysis. Am J Speech Lang Pathol. 2010;19(2):178–95. https://doi.org/10.1044/1058-0360(2010/09-0022).
61. Maglione MA, Gans D, Das L, et al. Nonmedical interventions for children with ASD: recommended guidelines and further research needs. Pediatrics. 2012;130(Suppl 2):S169–78. https://doi.org/10.1542/peds.2012-0900O.
62. Kasari C, Kaiser A, Goods K, et al. Communication interventions for minimally verbal children with autism: a sequential multiple assignment randomized trial. J Am Acad Child Adolesc Psychiatry. 2014;53(6):635–46. https://doi.org/10.1016/j.jaac.2014.01.019.

Hands-On Approach: A Novel Treatment for Autistic Children

5

Nidhi Sharma and Simranjeet Kaur

An extensive and thorough approach to autism rehabilitation includes a multidisciplinary, modified strategy that is based on the individual requirements and strengths of every person with autistic disorder. Such an approach typically involves various therapeutic modalities, educational interventions, and support services aimed at enhancing communication skills, social interaction, cognitive functioning, and adaptive behaviors. Here is an outline of key components that may be included in a comprehensive rehabilitation approach for autism:

5.1 Early Childhood Intervention Programs

Applied Behavioral Analysis Applied behavior analysis (ABA) stands as a beacon of scientific rigor and compassion, offering a transformative approach to understanding and shaping behavior, especially in individuals with developmental disabilities like autism spectrum disorder (ASD). At its core, ABA employs meticulous functional behavior assessments to unveil the intricate dynamics between antecedents, behaviors, and consequences, guiding practitioners to tailor interventions with precision. Through the strategic deployment of techniques such as positive reinforcement, shaping, prompting, and fading, ABA cultivates a pathway for individuals to acquire essential skills across domains including communication, social interaction, academic prowess, and daily living activities. Central to its success is the relentless pursuit of data-driven insights, enabling practitioners to track progress, refine strategies, and uphold accountability. ABA is not merely a therapeutic modality; it is a personalized journey, wherein the involvement of families and caregivers ensures consistency and fosters the seamless transfer of skills to diverse

N. Sharma (✉) · S. Kaur
Maharishi Markandeshwar Institute of Physiotherapy and Rehabilitation, Maharishi Markandeshwar (Deemed to be University), Mullana-Ambala, Haryana, India
e-mail: sharma.nidhi.physio@mmumullana.org; Simranjeet.kaur@mmumullana.org

M. Goyal, K. Goyal (eds.), *Rehabilitation Approach in Autism*,
https://doi.org/10.1007/978-981-96-4162-8_5

contexts. Despite historical critiques, modern ABA champions an ethos of dignity and empowerment, eschewing aversive methods in favor of positive reinforcement and ethical treatment. Its impact reverberates across educational, home, and clinical settings, offering hope and tangible results to countless individuals and families worldwide [1, 2].

Denver Model The Denver model is an early intervention program designed specifically for children between the ages of 18 months and 5 years who have ASD or related developmental disorder. It adopts a developmental approach, focusing on nurturing social, emotional, cognitive, and language development through naturalistic teaching strategies integrated into everyday activities and interactions, complemented by structured curriculum targeting specific skills. Central to the model is the active involvement of parents, who receive coaching to ensure consistency and application of skills at home and in community settings. To develop customized treatment programs that meet the requirements of each child, a multidisciplinary team of professionals—including psychologists, occupational therapists, speech-language pathologists, and special educators—works together. With a particular emphasis on communication, the Denver model utilizes techniques like modeling, prompting, and reinforcement to facilitate language development and enhance social communication skills. Supported by research, the Denver model has demonstrated positive outcomes in communication, social skills, and adaptive behavior, establishing itself as a research-based approach to early intervention for autistic children [1, 2].

Greenspan Method or DIR (Developmental Individual Differences Relationship-Based Floortime) Approach This therapeutic intervention framework primarily designed for a special child which is diagnosed with autism. It prioritizes the development of fundamental skills such as engagement, communication, and emotional regulation through child-directed, playful interactions that take place on the floor, hence the term "Floortime." Practitioners follow the child's lead, entering their world and encouraging reciprocal interactions while considering individual differences and developmental stages. Central to DIR Floortime is the recognition of relationships as pivotal to development, aiming to support children in reaching higher levels of functioning by addressing their unique developmental profiles. Through this child-centered approach, children are encouraged to explore and interact with their environment in a supportive and safe manner, fostering the acquisition of crucial social and emotional skills essential for lifelong learning and relationships [2].

5.2 Behavioral Therapies

Positive Behavior Support This is an evidence-based approach designed to enhance the way of life for individuals with developmental disabilities or behavioral disorders by reducing challenging behaviors. It prioritizes understanding the function or purpose of behavior and employing proactive strategies to prevent challenging behaviors while promoting positive alternatives. PBS strategies are individualized, focusing on teaching new skills, modifying environments, and providing positive reinforcement for desired behaviors. Key elements include conducting functional behavior assessments to identify triggers and maintaining factors, developing behavior intervention plans based on this assessment, and implementing proactive strategies to encourage positive behaviors. Collaboration among individuals, families, caregivers, and professionals is central to PBS, ensuring supportive environments and consistency across settings to improve the individual's overall well-being [2].

Cognitive-Behavioral Therapy A highly respected therapeutic technique recognizes the intricate connection linking thoughts, emotions, and behaviors. Its primary goal is to empower individuals by assisting them in recognizing and challenging negative or distorted thought patterns, leading to positive changes in behavior and emotions. CBT is characterized by its structured and goal-oriented nature, typically incorporating techniques such as cognitive restructuring, behavior activation, and problem-solving. Through collaborative efforts with a therapist, clients develop adaptive coping skills, gain insight into their thought processes, and learn practical strategies to manage challenges more effectively. CBT has been studied thoroughly and found to be beneficial for a number of mental health issues, such as PTSD, anxiety disorders, depression, and more. Its emphasis on skill-building and addressing the underlying causes of distress makes it a valuable tool for achieving lasting enhancements to general well-being and mental wellness [3–5].

Social Expertise Social expertise refers to the capacity to effectively navigate social interactions, demonstrating comprehension, empathy, and flexibility across diverse social contexts. It encompasses a broad spectrum of skills, including communication, emotional intelligence, perspective-taking, and conflict resolution. Individuals with social expertise can interpret social cues accurately, establish and maintain relationships, and handle social situations with confidence. They possess the ability to recognize and respond to the needs and emotions of others, cultivating positive connections and cooperation. Social expertise extends beyond mere social competence; it entails understanding the intricacies of social dynamics and cultural variations. It requires adapting behavior and communication styles to suit the situation and the people involved, fostering inclusivity and mutual understanding. Developing social expertise involves practice, introspection, and openness to learning and growth. Ultimately, strong social skills enhance personal and professional relationships, contributing to overall happiness and success in various aspects of life [6].

5.3 Speech-Language Therapy

Expressive Language Skills A vital intervention focuses on vocabulary growth, sentence structure, and articulation in order to help people with this disorder to enhance their expressive language skills. Therapists employ a range of techniques, such as modeling, prompting, and visual supports, to help individuals with autism communicate effectively. Through structured therapy sessions, therapists work with clients to build their expressive language skills, starting with basic vocabulary and sentence construction and progressing to more complex language concepts. Illustration supports, like face cards, communication boards, and electronic devices, are supplement verbal instruction for individuals with ASD with additional cues and prompts for communication. Therapists also incorporate play-based activities, interactive games, and role-playing exercises to engage individuals and practice communication skills in naturalistic settings. Additionally, therapists work closely with families and caregivers to reinforce therapy goals and strategies in everyday interactions at home and in the community. Speech and language therapy aims to improve not only communication abilities but also social interactions, academic performance, and overall well-being of life for individuals with ASD. It is the demand for therapy to be individualized to meet each person's unique needs and preferences, taking into account factors such as age, language level, and sensory sensitivities. Through consistent and structured intervention, individuals with autism can make significant progress in their expressive language skills and enhance their ability to interact and communicate with others effectively [1, 2, 5, 7].

Receptive Language Skills In addition to targeting expressive language skills, therapy also addresses language skills in which the autistic child has the ability to receive and focus on understanding spoken language, following directions, and interpreting nonverbal cues called receptive language. Therapists incorporate a variety of actions to improve comprehension and communication skills. It includes auditory processing tasks, such as listening to instructions or stories and answering questions, to enhance the ability to understand spoken language. Comprehension exercises involve activities like matching pictures to spoken words or sentences, identifying key information in a story, or answering questions about a passage. Interactive games and role-playing scenarios are used to practice interpreting non-verbal cue, for example, facial expressions, gestures, and body language, to comprehend the feelings and intentions of others. Additionally, therapists may use illustration supports, such as image schedule or societal stories, to reinforce concepts, which provide visual cue for understanding and following directions. It is essential for therapy to be individualized to meet each person's unique needs and preferences, taking into account factors such as age, language level, and sensory sensitivities. By targeting receptive language skills, therapy aims to progress overall communication ability, social interactions, and also academic performance of the autistic child, enhancing their ability to navigate and engage with the world around them effectively. Through consistent and structured intervention, individuals with autism can make significant progress in their receptive language skills and build a strong foundation for communication and social success [1, 2].

Pragmatic Language Skills Pragmatic language therapy is a specialized intervention that focuses on social communication skills, including turn taking, initiate conversations, and understanding social sign, in individuals with autism spectrum disorder (ASD). Therapists utilize various techniques such as role-playing, social scripts, and video modeling to teach and reinforce pragmatic language skills in real-life contexts. Role-playing activities allow individuals with ASD to practice social interactions, such as greeting others, asking questions, and expressing thoughts or feelings, in a structured and supportive environment. Social scripts provide individuals with pre-written phrases or scripts to use in specific social situations, helping them navigate social interactions more effectively. Watching movies of acceptable social behaviors and then replicating those actions in actual social settings is known as video modeling. Additionally, therapists may use visual supports, such as social stories or comic strip conversations, to help individuals with ASD understand social expectations and appropriate responses in various social situations. Pragmatic language therapy aims to improve social communication abilities, promote social interaction, and enhance overall social functioning for individuals with ASD. It is essential for therapy to be individualized to meet each person's unique needs and preferences, taking into account factors such as age, developmental level, and specific areas of difficulty. Through consistent and structured intervention, individuals with autism can make significant progress in their pragmatic language skills and develop more meaningful and fulfilling social relationships [2, 8].

5.4 Occupational Therapy

Sensory Integration The neurological system is responsible for organizing and interpreting sensory data from the outside world, giving people the ability to comprehend and react to their environment. It encompasses the integration of sensory inputs such as touch, taste, smell, sight, sound, and movement. Effective sensory integration allows individuals to filter out irrelevant sensory information and focus on what is important. However, individuals with sensory processing disorders may experience difficulties in sensory integration, impacting their ability to process and respond to sensory input. Occupational therapists often utilize sensory integration therapy to help individuals improve their sensory processing abilities. This therapy may involve activities designed to stimulate or calm the senses and strategies to assist individuals in modifying their responses to sensory input. By addressing sensory integration challenges, individuals can enhance their participation in daily activities and interactions, ultimately improving their overall quality of life [3, 9–11].

Motor Coordination Motor coordination refers to the skillful use of different body parts together to execute smooth and accurate movements. It encompasses the integration of sensory information, muscle strength, and motor planning abilities to perform various tasks effectively. Examples of motor coordination include walking, running, writing, and playing sports. Strong motor coordination is crucial for everyday activities, academic performance, and participation in recreational pursuits. Individuals with poor motor coordination may struggle with tasks requiring precise movements, such as tying shoelaces or using utensils. Occupational therapists and physical therapists often work with individuals to enhance motor coordination through tailored exercises and activities aimed at strengthening muscles, improving balance, and refining motor planning skills. By improving motor coordination, individuals can boost their independence, confidence, and overall quality of life [8].

Self-Care Skills Self-care skills are the fundamental activities individuals undertake to maintain their physical, emotional, and mental well-being on a daily basis. These encompass personal hygiene tasks like bathing, grooming, and dressing, as well as activities vital for health and well-being, such as eating balanced meals, exercising, and ensuring adequate rest. Additionally, self-care involves managing emotions and stress and employing relaxation techniques. These skills are crucial for fostering independence, autonomy, and overall well-being of life. In patients with disabilities or unique needs, self-care skills may require additional support and assistance from caregivers or professionals to develop and maintain. Occupational therapists often collaborate with individuals to enhance their self-care skills and devise strategies to promote independence in daily living activities. Mastering self-care skills contributes to improved self-esteem, confidence, and overall well-being [2, 4, 12].

5.5 Educational Support Services

Individualized Education Strategies (IES) Individualized education plans (IEPs) delineate educational objectives, adjustments, and supports for a child with autism, custom-tailored to meet their unique learning needs and foster academic success. These plans are crafted collaboratively by a team comprising educators, parents, and professionals such as special education teachers, therapists, and school psychologists. Following the SMART principles (specific, measurable, achievable, relevant, and time-bound), the IEP's objectives cover topics including behavior management, social skills, academic development, and communication.Adjustments and supports embedded in the IEP encompass specialized instruction, accommodations, and assistive technology to facilitate learning and active participation in the classroom. Additionally, strategies for sensory regulation, social integration, and emotional support are integrated to create a nurturing learning environment. Regular progress monitoring and review meetings ensure that the IEP remains responsive to the student's evolving needs and goals. Through personalized education plans, students with autism can receive the tailored support necessary to excel academically and achieve their full potential [12].

Assistive Technology Assistive technology tools, such as communication devices and visual supports, are instrumental in aiding students with autism spectrum disorder (ASD) to access the curriculum and communicate effectively. These tools are integrated into the classroom environment to provide support for learning and participation. Communication devices, like speech-generating devices or communication apps, empower students with limited verbal skills to express themselves, engage in classroom activities, and interact with peers and teachers. Meanwhile, visual supports such as visual schedules, picture cards, and social stories help students understand expectations, organize their day, and navigate social interactions. By offering visual cues, structure, and reinforcement, these tools help students with ASD succeed academically and socially. Moreover, assistive technology fosters independence, improves attention and focus, and reduces anxiety, enabling students to thrive in various educational settings. Through collaboration between teachers and therapists, assistive technology is selected and implemented to meet the individual needs of each student, ensuring equitable access to resources and supporting them in achieving their educational goals. By breaking down barriers to learning and communication, assistive technology empowers students with autism to reach their full potential [3, 6, 7].

Special Education Teachers and Aides Special education teachers and paraprofessionals play vital roles in providing tailored instruction, support, and accommodations to students with autism to ensure their academic and social success. These experts work together with the student's individualized education plan (IEP) team to create and carry out customized lesson plans that are tailored to meet each student's specific needs. They adeptly adapt curriculum materials, modify teaching strategies, and provide necessary accommodations to make learning accessible and meaningful for children with ASD. In addition, special education teachers and paraprofessionals are skilled in implementing behavior management techniques, often devising behavior intervention plans to address challenging behaviors and foster positive social interactions. They also offer individualized support and guidance to help students navigate social situations, develop essential social skills, and cultivate positive relationships with peers and adults. Furthermore, they work collaboratively with families to ensure consistent support across home and school settings. Through their dedication and expertise, special education teachers and paraprofessionals create inclusive learning environments that empower students with autism to thrive academically, socially, and emotionally [13].

5.6 Parent Training and Support

Education and Skill Building Parent training programs provide essential information, resources, and support to help parents better understand autism and learn effective strategies for supporting their child's development. These programs cover a wide range of topics, including behavior management techniques, communication strategies, and advocacy skills. In behavior management, parents learn about research-based approaches like positive reinforcement, visual supports, and structured routines to address challenging behaviors effectively. Communication techniques focus on teaching parents how to facilitate language development, improve social skills, and support their child's communication needs through methods such as augmentative and alternative communication (AAC) tools or image exchange communication systems (PECS). Moreover, parent training programs equip parents with advocacy skills to navigate the complex educational and healthcare systems, ensuring their child receives appropriate services and accommodations. By participating in these programs, parents gain confidence, knowledge, and practical skills to support their child's growth and development effectively. Additionally, they benefit from peer support and networking opportunities, fostering a sense of community and empowerment. Overall, parent training programs play a crucial role in empowering parents to become effective advocates and partners in their child's journey with autism [5, 14, 15].

Behavior Management Techniques Parents of children with ASD can get evidence-based behavior management skills in parent support groups, in addition to receiving emotional support. These approaches can help parents handle troublesome behaviors and encourage good alternatives for their children. These groups often offer workshops or training sessions where parents can receive instruction in techniques such as reinforcement, visual supports, and structured routines. Through these training sessions, parents gain valuable insights into understanding the underlying causes of their child's behavior and learn effective strategies to address them. Reinforcement techniques help parents understand how to reinforce desired behaviors while minimizing or redirecting challenging ones. Visual supports, such as visual schedules or picture cards, provide a visual structure that can help children understand routines and expectations, reducing anxiety and promoting independence. Additionally, learning about structured routines helps parents establish predictable schedules and environments, which can be comforting and reassuring for children with ASD. By equipping parents with these evidence-based techniques, support groups empower them to effectively manage their child's behavior and create a positive and supportive home environment conducive to their child's growth and development [1, 2, 4, 10].

Emotional Support and Coping Strategies Parent support groups play an important role in providing a supportive surrounding for parents of children with autism spectrum disorder (ASD) to connect, share experiences, and access emotional support. These groups serve as a forum for parents to discuss challenges, celebrate successes, and seek advice from others who understand their experiences firsthand. By connecting with other parents facing similar circumstances, individuals can gain valuable insights, resources, and encouragement. Additionally, parent support groups provide opportunities for parents to learn from one another and develop effective coping strategies for managing stress and uncertainty. Through group discussions, workshops, and guest speakers, parents can access information on various topics related to ASD, such as behavior management, educational resources, and accessing services. Additionally, these groups frequently promote a sense of belonging and community, which lessens feelings of loneliness and gives families access to a network of support. Parent support groups enable parents to successfully manage the difficulties of raising a child with ASD and advocate for their kid's needs by providing empathy, validation, and helpful advice [5, 10–12, 14, 16].

5.7 Counseling and Mental Health Services

Individual Therapy People with autism spectrum disorder (ASD) can explore their thoughts, feelings, and experiences in a safe and encouraging setting through individual treatment. Therapists collaborate closely with patients to address issues pertaining to mental health and autism, improve self-awareness, and create coping mechanisms throughout therapy sessions. Therapists can employ a range of methods, depending on the requirements and preferences of the patient, including mindfulness-based approaches, cognitive-behavioral therapy (CBT), and psychoeducation. Through individual therapy, clients with ASD can learn to identify and regulate their emotions, improve communication skills, and develop effective problem-solving strategies. Additionally, therapists provide guidance and support as clients navigate social interactions, relationships, and transitions. By fostering a trusting and empathetic therapeutic relationship, individuals with ASD can gain a deeper understanding of themselves and their experiences, leading to increased self-confidence and improved overall health. It is important for therapists to have specialized training and experience in working with individuals with ASD to provide effective and culturally sensitive care. Through individual therapy, individuals with autism can build resilience, improve coping skills, and achieve greater personal growth and fulfillment [12].

Family Treatment Family therapy is an essential approach that works collaboratively with families to improve communication, manage issues, and strengthen relationships. Therapists in family treatment help families understand the impact of autism on family dynamics and develop strategies to support each other and promote positive interactions. Through family therapy sessions, therapists create a supportive and nonjudgmental space where family members can express their concerns, share their experiences, and work together to address challenges related to autism.

Therapists provide psychoeducation about autism spectrum disorder (ASD), helping family members gain a better understanding of the condition and its effects on behavior, communication, and social interactions. They also teach effective communication and problem-solving skills, enabling family members to express their needs, resolve conflicts, and build stronger connections. Additionally, therapists help families navigate transitions, such as changes in routines or roles, and provide support during times of stress or crisis. When the whole family participates in therapy, loved ones consistently provide support and encouragement to those with ASD, which improves outcomes and enhances everyone's quality of life. It is important for therapists to recognize and respect the unique cultural and familial dynamics of each family, adapting interventions to meet their specific needs and preferences. Through family treatment, families can learn to navigate the challenges of autism together, building resilience and fostering healthier relationships [2–4, 10].

Psychopharmacological Treatments Psychopharmacological interventions play a significant role in addressing co-occurring mental health disorders, such as stress or depression, in people with autism. These interventions involve the use of medications prescribed and monitored by a psychiatrist to manage symptoms and improve overall functioning. For instance, doctors may give selective serotonin reuptake inhibitors (SSRIs) to treat ASD patients' anxiety or depression symptoms. Antipsychotic drugs can also be used to treat symptoms including irritability, violence, or compulsive behavior. Methylphenidate and other amphetamine-based medications are examples of stimulant medications that may be administered to treat attention deficit hyperactivity disorder (ADHD), a condition that is frequently observed in people with ASD. Additionally, mood stabilizers or anticonvulsants may be used to manage mood swings or aggression in some cases. It is important to communicate that the use of psychopharmacological interventions in ASD is carefully monitored, and medication decisions are made on an individual basis, taking into account factors such as age, co-occurring conditions, and potential side effects. Psychiatrists work personally with the person and their family to determine the most appropriate medication regimen, monitor treatment progress, and adjust dosages as needed. To provide complete support for people with ASD and co-occurring mental health illnesses, psychopharmacological interventions are frequently utilized in conjunction with other therapies, such as behavioral interventions or psychotherapy [5, 9, 16].

5.8 Transition Planning and Vocational Support

Transition Assessments Transition planning is a crucial process that begins in adolescence and focuses on preparing the person with autism spectrum disorder for the transition to adulthood. This comprehensive planning involves assessing the person's strengths, benefit, and support needs to develop a personalized transition plan. Assessments may include vocational evaluations, which help identify the individual's skills, interests, and abilities related to potential career paths. Functional assessments evaluate the individual's daily living skills, such as managing money, using public transportation, and independent living skills. Additionally, transition planning meetings bring together the individual, their family, educators, and other professionals to discuss goals, resources, and support services needed during the transition period. These meetings aim to create a roadmap for the individual's transition to adulthood, outlining goals and steps to achieve them. Transition planning also includes exploring postsecondary education options, vocational training, employment opportunities, and community resources. By engaging in transition planning, individuals with ASD can develop the necessary skills and supports to navigate the challenges of adulthood successfully [6, 12, 17].

Postsecondary Education Supporting people with autism spectrum disorder (ASD) during their transition to postsecondary education options, like universities or vocational training programs, is a critical function of transition services. In order to guarantee that students with ASD have access to the tools and accommodations they require to succeed in higher education, these programs include a variety of assistance. Academic accommodations may include specialized instruction, extended time on exams, note-taking assistance, and alternative formats for assignments. Support services, such as counseling, tutoring, and peer mentoring, provide additional assistance to help students navigate academic and social challenges. Transition programs specifically tailored for students with ASD offer workshops, support groups, and individualized coaching to help them develop essential skills for college success, such as time management, organization, and self-advocacy. These programs also provide opportunities for students to connect with peers, build social networks, and access community resources. By providing comprehensive transition services, individuals with ASD can successfully transition to postsecondary education, achieve their academic goals, and prepare for meaningful careers and independent living [10, 12, 17].

Employment Services Vocational training programs and supported employment services are essential for individuals with autism to develop job skill, explore career options, and find meaningful employment opportunities. These programs offer specialized training and support to help individuals with ASD acquire the skills and confidence needed to succeed in the workplace. Vocational training programs provide hands-on instruction in specific job skills, such as computer programming, culinary arts, or automotive repair, tailored to the individual's interests and abilities. Supported employment services offer job coaching and on-the-job support to individuals with ASD as they navigate the job search process and transition into employment. Job coaches provide assistance with tasks such as resume writing, job searching, and interview preparation, helping individuals with ASD present their skills and qualifications effectively to potential employers. Additionally, job coaches provide ongoing support in the workplace, helping individuals with ASD learn job tasks, communicate effectively with coworkers and supervisors, and navigate workplace social dynamics. By participating in vocational training programs and supported employment services, individuals with ASD can gain valuable job skills, build confidence, and achieve greater independence and self-sufficiency in the workforce [2, 4, 6, 12, 13].

Independent Living Skills For people with autism spectrum disorder (ASD), transition planning is essential in helping them acquire the independent living skills needed for adult autonomy. Finding and developing skills like budgeting, using public transit, and upkeep of a household are all part of this process. Training programs tailored for individuals with ASD provide instruction and practice in these essential life skills. These programs may offer workshops, hands-on activities, and real-world simulations to teach budgeting, shopping, cooking, and household management. Supported living arrangements give opportunity for individuals with ASD to practice and apply this skill in a supportive environment. These arrangements may include group homes, supervised apartments, or independent living with assistance from support staff or caregivers. In these settings, individuals receive guidance and support as they learn to manage their daily routines, make decisions, and solve problems independently. By participating in transition planning and accessing training programs and supported living arrangements, individuals with ASD can develop the skill and self-confidence required for successful independent living in adulthood [2, 12, 15, 18].

5.9 Community Integration and Inclusion Initiatives

Recreational Activities Recreational programs that are centered in the community are essential in giving people with autism spectrum disorder (ASD) the chance to socialize, make friends, and participate in leisure activities. These programs provide a range of activities catered to the interests and requirements of people with ASD, including sports teams, art workshops, and social organizations. Sports teams provide opportunities for individuals to participate in team sports, improving physical fitness and coordination while fostering teamwork and social interaction. Art classes offer a creative outlet for self-expression and can help individuals develop fine motor skills and confidence in their abilities. Social clubs provide structured social opportunities where individuals can practice social skills, learn to initiate and maintain conversations, and build friendships in a supportive environment. Additionally, community-based recreational programs often incorporate sensory-friendly environments and accommodations to ensure that individuals with ASD feel comfortable and included. By participating in these programs, individuals with autism can enhance their quality of life, increase social connectedness, and develop the skills necessary for meaningful participation in their communities [2, 6].

Community Events and Activities Access to community events and activities plays a vital role in promoting social inclusion and participation for people with autism spectrum disorder (ASD). These events, which may include festivals, concerts, or community service projects, offer opportunities for individuals with ASD to connect with others and contribute to their communities in meaningful ways. Festivals and concerts provide settings where individuals with ASD can enjoy sensory-rich experiences while being part of a larger social environment. These events often offer sensory accommodations such as quiet zones or sensory-friendly activities to ensure that individuals with ASD feel comfortable and included. Community service projects allow individuals with autism to engage in activities that benefit others and foster a sense of purpose and accomplishment. These projects may involve volunteering at local charities, participating in environmental cleanup efforts, or organizing fundraising events. By participating in community events and activities, individuals with autism can develop social skills, expand their social networks, and build confidence in their abilities to interact with others. Additionally, these experiences promote understanding and acceptance of individuals with ASD within the broader community, fostering a more inclusive society for all [1, 2, 12].

Advocacy and Awareness Initiatives Autism societies and advocacy groups are essential in advancing acceptance, increasing knowledge, and fighting for the rights of people with autism spectrum disorder (ASD). These organizations assist people with ASD and their families in navigating obstacles and obtaining necessary services by offering a wide range of resources, support, and education. Furthermore, advocacy groups use community gatherings, educational programs, and public awareness campaigns to advance societal acceptance and understanding of autism. Additionally, they push for federal, state, and local policy changes to guarantee that people with ASD have access to the right kind of support services, education, and medical treatment. Additionally, these groups support social inclusion programs that seek to make communities, businesses, and educational institutions more accepting of people with autism. By raising awareness, providing support, and advocating for change, advocacy organizations and autism societies empower individuals with ASD to lead fulfilling lives and participate fully in society [1, 12, 13].

5.10 Holistic and Complementary Therapies

Music Therapy Music therapy is a valuable management for people with autism spectrum disorder (ASD), utilizing vocal-based techniques such as singing, playing instruments, listening to music, and improvisation to address various needs, including communication, social skills, emotional regulation, and sensory integration. Through structured music activities, therapists tailor interventions to the different needs and preferences of every individual, promoting self-expression, emotional regulation, social engagement, and sensory integration. For communication, music therapy may involve singing or playing instruments to facilitate verbal expression or using rhythm and melody to enhance receptive language skills. In social contexts, group music-making activities encourage interaction, turn-taking, and cooperation among participants. Music therapy also supports emotional regulation by providing opportunities for individuals to express and process emotions through music. Additionally, music activities can help individuals with ASD manage sensory sensitivities by providing predictable rhythms, patterns, and sensory input in a controlled environment. Overall, music therapy enhances the overall health and quality of life for people with autism by providing a creative and engaging means of addressing a range of needs and promoting holistic development [3, 7, 10, 14].

Art Therapy Art therapy is a potent strategy that supports people with autism spectrum disorder (ASD) in many areas of their development by encouraging artistic expression and engagement in artistic activities. Individuals with ASD can explore their thoughts, feelings, and sensory experiences in a safe and encouraging atmosphere through art therapy activities like painting, drawing, sculpting, and collage making. By engaging in these creative activities, individuals can enhance communication skills, express feelings that may be difficult to articulate verbally, and process complex emotions. Art therapy also promotes relaxation and stress reduction, helping individuals to manage anxiety and improve overall well-being. Furthermore, the process of creating art can increase self-awareness and self-esteem, as individuals see their ideas and emotions take form in their artwork. Additionally, as people gain understanding of their experiences and create coping mechanisms, art therapy can support emotional recovery and give them a sense of empowerment. Overall, art therapy offers a unique and effective approach to supporting individuals with autism, providing an alternative means of communication and self-discovery that promotes holistic well-being [11].

Animal-Assisted Rehabilitation Animal-assisted rehabilitation involves interactions with specially trained animals, such as dogs, horses, or dolphins, to enhance physical, emotional, and social well-being in people with autism. These interactions, often facilitated by trained therapists or handlers, provide opportunity for people with ASD to engage in activities such as petting, grooming, walking, and playing with animals. These activities can have profound effects on individuals with ASD, reducing stress, anxiety, and sensory sensitivities while fostering companionship, empathy, and social skills. Interacting with animals can provide a calming effect, as the rhythmic movements and predictable behavior of animals can help regulate emotions and decrease arousal levels. In addition, people with ASD can practice social interactions and communication skills in a safe and encouraging atmosphere thanks to the nonjudgmental nature of animals. Animal-assisted activities can also encourage physical activity and motor skills development through activities like walking or grooming animals. All things considered, animal-assisted rehabilitation provides a comprehensive therapeutic strategy that attends to the physical, emotional, and social needs of people with ASD, enhancing quality of life and fostering well-being [19–21].

Yoga and Mindfulness Practices Yoga and mindfulness-based interventions are effective approaches that integrate breathing exercises, meditation, relaxation techniques, and gentle physical postures to maintain people with autism spectrum disorder in various aspects of their well-being. These practices aim to promote self-regulation, stress reduction, body awareness, and emotional resilience. Through yoga and mindfulness, individuals with ASD can learn to connect with their bodies, regulate their breathing, and cultivate a sense of calmness and relaxation. People who practice mindfulness are better able to control their stress and anxiety because they are more conscious of their thoughts, feelings, and physical sensations. Moreover, yoga and mindfulness activities enhance concentration, focus, and emotional self-awareness, providing valuable coping strategies for managing challenges such as sensory overload and emotional dysregulation. The gentle physical postures practiced in yoga can also improve flexibility, balance, and coordination, contributing to overall physical well-being. Overall, yoga and mindfulness-based interventions offer individuals with ASD practical tools for self-care and emotional regulation, empowering them to navigate daily challenges with greater ease and resilience [22–26].

Dietary Interventions Dietary interventions encompass various modifications to the diet aimed at addressing nutritional deficiencies, gastrointestinal issues, and sensory sensitivities in people with autism spectrum disorder. These modifications may comprise by eliminating certain foods, additives, or allergens, as well as incorporating nutrient-rich foods, supplements, or specialized diets such as gluten-free or casein-free. While the evidence supporting the efficacy of dietary interventions in autism is mixed, some individuals may benefit from personalized dietary approaches tailored to their specific needs and preferences. For example, some research suggests that certain dietary changes, such as reducing gluten or casein intake, may alleviate gastrointestinal symptoms and improve behavior in some individuals with ASD. Additionally, nutrient supplementation, such as omega-3 fatty acids or vitamins, may help address nutritional deficiencies commonly observed in individuals with ASD. Furthermore, embracing a varied and well-balanced diet high in fruits, vegetables, whole grains, and lean proteins can enhance general health and well-being by offering vital nutrients. It is critical that families and people with ASD collaborate with medical specialists, such as nutritionists or registered dietitians, to create individualized, safe, and sustainable food programs. By making informed dietary choices, individuals with ASD can potentially improve their overall health and quality of life [27, 28].

5.11 Conclusion

A comprehensive rehabilitation approach for autism encompasses a wide range of literature-based treatment and supportive services adapted to the exceptional needs, strengths, and developmental stage of each individual with autism. Collaboration among professionals, families, educators, and community stakeholders is essential to ensure coordinated and holistic support across various domains of functioning, promoting optimal outcomes and the standard of living for people with autism and their families.

References

1. Roane HS, Fisher WW, Carr JE. Applied behavior analysis as treatment for autism spectrum disorder. J Pediatr [Internet]. 2016;175:27–32. https://doi.org/10.1016/j.jpeds.2016.04.023.
2. Vismara LA, Rogers SJ. Behavioral treatments in autism spectrum disorder: what do we know? Annu Rev Clin Psychol. 2010;6:447–68. https://doi.org/10.1146/annurev.clinpsy.121208.131151.
3. Moossavi A, Moallemi M. Auditory processing and auditory rehabilitation approaches in autism. Audit Vestib Res. 2019;28(1):1–13. https://doi.org/10.18502/avr.v28i1.410.
4. Dubreucq M, Dubreucq J. Toward a gender-sensitive approach of psychiatric rehabilitation in autism spectrum disorder (ASD): a systematic review of women needs in the domains of romantic relationships and reproductive health. Front Psychiatry. 2021;12:630029. https://doi.org/10.3389/fpsyt.2021.630029.
5. DeFilippis M, Wagner KD. Treatment of autism spectrum disorder in children and adolescents. Psychopharmacol Bull. 2016;46(2):18–41.
6. Bozgeyikli LAL, Bozgeyikli E, Raij A, Alqasemi R, Katkoori S, Dubey R. Vocational rehabilitation of individuals with autism spectrum disorder with virtual reality. ACM Trans Access Comput. 2017;10(2):5.1–25. https://doi.org/10.1145/3046786.
7. Golestan S, Soleiman P, Moradi H. A comprehensive review of technologies used for screening, assessment , and rehabilitation of autism spectrum disorder. ArXiv. 2018;abs/1807.1:1–12.
8. Martinez E, Garcia A. Virtual game approach for rehabilitation in autistic children. In: 2011 8th international conference on information, communications & signal processing. Singapore; 2011. p. 1–6. https://doi.org/10.1109/ICICS.2011.6174256.
9. Shenoy MD, Indla V, Reddy H. Comprehensive management of autism: current evidence. Indian J Psychol Med. 2017;39(6):727–31. https://doi.org/10.4103/IJPSYM.IJPSYM_272_17.
10. Frye RE. A personalized multidisciplinary approach to evaluating and treating autism spectrum disorder. J Pers Med. 2022;12(3):464. https://doi.org/10.3390/jpm12030464.
11. Wang M, Reid D. Using the virtual reality-cognitive rehabilitation approach to improve contextual processing in children with autism. Sci World J. 2013;2013:716890. https://doi.org/10.1155/2013/716890.
12. Sansosti FJ, Merchant D, Koch LC, Rumrill P, Herrera A. Providing supportive transition services to individuals with autism spectrum disorder: considerations for vocational rehabilitation professionals. J Vocat Rehabil. 2017;47(1):207–22. https://doi.org/10.3233/JVR-170896.
13. Shattuck PT, Grosse SD. Issues related to the diagnosis and treatment of autism spectrum disorders. Ment Retard Dev Disabil Res Rev. 2007;13(2):129–35. https://doi.org/10.1002/mrdd.20143.
14. Nazam F, Husain A. Cognitive impairment and rehabilitation of children and adults with autism spectrum disorder. In: Md Ashraf G, Alexiou A, editors. Autism spectrum disorder and Alzheimer's disease [Internet]. Singapore: Springer Nature; 2022. p. 301–14. https://doi.org/10.1007/978-981-16-4558-7_15.

15. Wickstrom J, Dell'Armo K, Salzman E, Hooker JL, Delehanty A, Bishop S, Tassé MJ, Wetherby AM, Piergies AMH, Damiano D, Rauch A, Thurm A. Systematic review: recommendations for rehabilitation in ASD and ID from clinical practice guidelines. Arch Rehabil Res Clin Transl. 2021;3(3):100140. https://doi.org/10.1016/j.arrct.2021.100140.
16. Calderoni S, Billeci L, Narzisi A, Brambilla P, Retico A, Muratori F. Rehabilitative interventions and brain plasticity in autism spectrum disorders: focus on MRI-based studies. Front Neurosci. 2016;10:139. https://doi.org/10.3389/fnins.2016.00139.
17. Chang YC, Locke J. A systematic review of peer-mediated interventions for children with autism spectrum disorder. Res Autism Spectr Disord. 2016;27:1–10. https://doi.org/10.1016/j.rasd.2016.03.010.
18. Lang R, Regester A, Rispoli M, Pimentel S, Camargo H. Rehabilitation issues in autism spectrum disorders. Dev Neurorehabil. 2010;13(3):153–5. https://doi.org/10.3109/17518421003607597.
19. Ang CS, MacDougall FA. An evaluation of animal-assisted therapy for autism spectrum disorders: therapist and parent perspectives. Psychol Stud (Mysore) [Internet]. 2022;67(1):72–81. https://doi.org/10.1007/s12646-022-00647-w.
20. Koukourikos K, Georgopoulou A, Kourkouta L, Tsaloglidou A. Benefits of animal assisted therapy in mental health. Int J Caring Sci [Internet]. 2019;12(3):1898. Available from: www.internationaljournalofcaringsciences.org
21. Bert F, Gualano MR, Camussi E, Pieve G, Voglino G, Siliquini R. Animal assisted intervention: a systematic review of benefits and risks. Eur J Integr Med. 2016;8(5):695–706. https://doi.org/10.1016/j.eujim.2016.05.005.
22. Singh NN, Lancioni GE, Karazsia BT, Myers RE, Hwang YS, Anālayo B. Effects of mindfulness-based positive behavior support (MBPBS) training are equally beneficial for mothers and their children with autism spectrum disorder or with intellectual disabilities. Front Psychol. 2019;10:385. https://doi.org/10.3389/fpsyg.2019.00385.
23. Ridderinkhof A, de Bruin EI, Blom R, Bögels SM. Mindfulness-based program for children with autism spectrum disorder and their parents: direct and long-term improvements. Mindfulness (N Y). 2018;9(3):773–91. https://doi.org/10.1007/s12671-017-0815-x.
24. Sun Y, Lamoreau R, O'Connell S, Horlick R, Bazzano AN. Yoga and mindfulness interventions for preschool-aged children in educational settings: a systematic review. Int J Environ Res Public Health. 2021;18(11):6091. https://doi.org/10.3390/ijerph18116091.
25. Radhakrishna S, Nagarathna R, Nagendra HR. Integrated approach to yoga therapy and autism spectrum disorders. J Ayurveda Integr Med. 2010;1(2):120–4. https://doi.org/10.4103/0975-9476.65089.
26. Semple RJ. Review: yoga and mindfulness for youth with autism spectrum disorder: review of the current evidence. Child Adolesc Mental Health. 2019;24(1):12–8. https://doi.org/10.1111/camh.12295.
27. Baspinar B, Yardimci H. Gluten-free casein-free diet for autism spectrum disorders: can it be effective in solving behavioural and gastrointestinal problems? Eurasian J Med. 2020;52(3):292–7. https://doi.org/10.5152/eurasianjmed.2020.19230.
28. Whiteley P, Shattock P, Knivsberg AM, Seim A, Reichelt KL, Todd L, Carr K, Hooper M. Gluten- and casein-free dietary intervention for autism spectrum conditions. Front Hum Neurosci. 2013;6:344. https://doi.org/10.3389/fnhum.2012.00344.

Occupational Therapy for Autistic Children

6

Gargi Mishra and Durga Prasad Mishra

6.1 Learning Objectives

This chapter covers:

1. *Understanding Autism:* Here, readers will learn what autism spectrum disorder (ASD) is. This means knowing its definition, features, and how it is diagnosed. Readers should get a good sense of how ASD affects people's lives in various ways.
2. *Incidence and Prevalence of Autism:* In this section, we focus on the statistics around autism spectrum disorder. Readers will learn about how common ASD is—globally and within certain groups. Also included are trends or patterns regarding its frequency.
3. *Understanding the Role of Occupational Therapy:* This part aims to clarify what occupational therapy does for those with ASD. Readers need to grasp the main ideas and methods that guide OTs in treating autism.
4. *Application of Occupational Therapy Interventions:* After understanding the OTs' role, readers should see how to put this knowledge into practice with individuals who have autism spectrum disorder. This includes learning how OTs evaluate needs and create treatments to meet those specific needs.
5. *Identification and Description of Therapeutic Approaches:* Here we explore the different treatment methods OTs frequently use when working with individuals on the spectrum. Readers must be able to recognize and describe these methods along with their core principles.
6. *Analysis of Therapeutic Approaches from an Evidence-Based Practice Perspective:* This section highlights why evidence-based practice matters in

G. Mishra
Utkal University, Bhubaneswar, Odisha, India

D. P. Mishra (✉)
School of Occupational and Physiotherapy, Driems University, Cuttack, Odisha, India

M. Goyal, K. Goyal (eds.), *Rehabilitation Approach in Autism*,
https://doi.org/10.1007/978-981-96-4162-8_6

occupational therapy for ASD. Readers should critically assess popular therapies by looking at their effectiveness using research evidence and actual results.

By achieving these learning goals, readers will gain a thorough understanding of ASD—how common it is—and the different evidence-backed treatment options available to support those living with it. They will also learn how occupational therapy plays a part in treating and supporting individuals with ASD and how to use evidence-based practices to improve results.

6.2 Background Information

A child with autism may find comfort in the chaos of recess. Spinning in circles can provide a sense of predictability. This simple movement becomes calming. However, certain fabrics can be a challenge. The child might cringe at how they feel against their skin. Their hypersensitive senses react strongly to different textures.

Eating can also be a struggle. For many, the sensation of some foods their mouth can feel overwhelming. Sensory sensitivities often dictate what they can tolerate. When frustration reaches its limit, communication barriers become evident. This can result in explosive outbursts. Expressing needs and emotions is difficult for them. The world surrounding them may seem confusing and overwhelming at times.

ASD cases have increased significantly over the past decade, according to CDC data. This rise underscores the crucial demand for effective services as well as agendas to backing persons with ASD afterward their families in numerous aspects of life, together with education, family dynamics, and employment.

Occupational therapy (OT) practitioners play a significant position in designing plus delivering interventions aimed at individuals having ASD. Their focus is on promoting purposeful performance with participation in day-to-day activities, enhancing engagement with the environment, and fostering social integration [1].

To ensure the most effective interventions, Occupational therapy need to stay informed about evidence-based practices. This includes understanding the latest research findings and employing interventions that have established effectiveness in improving outcomes for persons with ASD.

Almost all children with ASD may require occupational therapy and many can benefit greatly from it. Occupational therapy can help children along with adolescents having ASD develop skills toward navigating daily activities, interacting with peers, and achieving greater independence. By tailoring interventions to meet specific needs individually, occupational therapy contribute significantly to enhancing the quality of life meant for those having ASD with their families.

6.3 Understanding Autism Spectrum Disorder

6.3.1 Definition of Autism Spectrum Disorder

Autism is a developmental condition, which is characterized primarily through behavioural traits. These encompass challenges in speech and language acquisition, social interaction, play, and behaviour [2]. Autism spectrum disorder's diagnostic criteria are outlined in DSM-5: (a) Social communication and interaction are consistently challenging in various situations. (b) Engaging in repetitive behaviours, exhibiting restricted interests, or adhering to rigid routines. (c) Early development may show symptoms, but they may not be fully visible until faced with increasing social demands or may be hidden by learned coping mechanisms later in life. (d) The functioning of social, occupational, or other significant aspects of daily life can be impeded by these symptoms. (e) Intellectual disability or global developmental delay cannot be the sole cause of the observed disturbances.

6.3.2 Incidence and Prevalence of ASD

It is indeed concerning to see the rising prevalence of ASD globally [3], including in India. Different regions and studies have different ASD prevalence rates due to different factors such as differences in screening methods, diagnostic criteria, and population demographics. In India, recent studies have reported a prevalence range of 0.15–1.01% [4, 5], indicating the variability in estimates. Increased awareness among professionals and changes in diagnostic criteria certainly contribute to these numbers; besides factors such as perinatal risk factors and advanced parental age at conception are also considered.

6.3.3 Characteristics of ASD

The symptoms of ASD often show up early, generally becoming clear around the age of 18 months. Most diagnoses happen between 3 and 5 years old in countries with good resources [6].

Kids with ASD may act differently than other children. They find it hard to interact socially. For example, they might not make eye contact or respond when their name is called. They may engage in repetitive behaviours like hand flapping and repeating the same words or actions [7].

Many children with ASD found difficulties while initiating and maintaining communication. They may have delays in developing speech, and some might repeat words or phrases they have heard. Besides they might react unusually to various sensory experiences

Many children with ASD face challenges in communication. They often experience delays in learning to talk, and some may engage in echolalia—this means they repeat words or phrases they hear. Understanding and expressing feelings can be tough for them too. Additionally, they might react strangely to different sensory experiences.

Another key trait of ASD is a strong need for routine. Changes to their schedule can be hard for them to handle, sometimes leading to tantrums when things do not go as expected.

Additionally, certain children with ASD may suffer regression, losing previously attained skills or words [7].

6.3.4 Diagnosis of ASD

The DSM-5, the diagnostic manual used by clinicians, categorizes indications of ASD into two main dominions:

(A) *Social communication and interaction*: Both verbal and non-verbal communication skills can be identified as deficits in this domain. Some examples include difficulties with reciprocal conversation; challenges in understanding and using non-verbal cues, for instance, facial expressions with gestures; and lack of typical social reciprocity.

(B) *Restricted and repetitive behaviours*: This domain involves repetitive behaviours, stereotypical movements, insistence on consistency or routines, and highly concentrated interests that are abnormal in intensity or focus.

Clinicians use these criteria, along with other relevant information such as developmental history and observation, to make a diagnosis of ASD (Table 6.1). The DSM-5 provides specific criteria within each domain that clinicians use to assess whether an individual satisfies the ASD criteria [2].

To be diagnosed with ASD, symptoms must be present from early childhood according to DSM-5 criteria and typically result in significant impairment in daily functioning. This means that the symptoms and behaviours associated with ASD significantly impact a person's capability to effectively navigate social interactions, communicate, engage in typical activities, and adapt to change. The presence of these symptoms and their impact on everyday life is crucial for determining and diagnosing ASD.

Severity levels also need to be addressed in the DSM-5 (Table 6.2). The severity levels are specified to reflect the degree of social communication impairments and restricted stereotyped behaviours. They are categorized into three levels:

1. *Level 1 (requiring support):* Social communication impairments that are noticeable require support for individuals with Level 1 ASD. Some people might find it hard to start talking to others. They could act differently or struggle when trying to respond. Plus, the way they repeat certain actions can get in the way of doing well in different situations.
2. *Level 2 (requiring substantial support):* When it comes to Level 2 ASD, people need quite a bit of support for social communication. Starting chats can be tough

Table 6.1 DSM-5 criteria for autism

Social communication and interaction:	Repetitive and restrictive behaviours:
1. *Problems with social reciprocity:* Unusual ways of approaching others Not responding normally in social situations Not sharing interests, emotions, or reactions with others No spontaneous social interactions 2. *Issues with non-verbal communication:* Difficulty combining spoken language and gestures Strange eye contact and body movements Trouble understanding and using non-verbal cues Lack of facial expressions and gestures 3. *Challenges in relationships:* Hard time making friends Little interest in people Trouble adjusting behaviour to different situations	1. *Repetitive speech and actions:* Using the words or phrases repeatedly Repeating simple movements or actions Using objects in repetitive ways Using phrases that are unusual 2. *Sticking to routines:* Strong resistance to routine changes, e.g. exercise habits Wanting the same routes and diet always Extreme stress or repeated questioning over small changes 3. *Intense interests:* Strong attachment to unusual objects Narrow and intense interests 4. *Unusual reactions to sensory input:* Odd interest in sensory experiences (e.g., lights, textures) Indifference to pain or temperature changes Negative reactions to certain sounds or textures Excessive touching or sniffing of objects

Content adapted from DSM-5 [2]

Table 6.2 The severity levels of autism

Severity levels	Level 3	Level 2	Level 1
Support required by the child	Very substantial	Substantial	Required
Lack of verbal communication skills	Severe	Marked	Noticeable when no support in place
Non-verbal communication skills deficits	Severe	Marked	Noticeable when no support in place
Impairments in social functioning	Severe	Apparent even with supports in place	Decreased interest in social interaction
Initiation of social interaction	Very limited	Limited	Difficult, odd, or unsuccessful
Response to social overtures from others	Minimal	Reduced	Atypical or unsuccessful
Behavioural inflexibility	Present	Present	Interferes with functioning
Coping to change	Extremely difficult	Difficult	Problems of organization and planning
Restricted/repetitive behaviours	Markedly affect function	Interfere with function	
Changing focus or action	Causes great distress	Difficult/causes distress	Difficult

Content adapted from DSM-5 [2]

for them. They might only react to very direct and clear social cues. Also, their repetitive actions can be more noticeable, really affecting how they do things in daily life.

3. *Level 3 (requiring very substantial support):* Now, with Level 3 ASD, social communication issues are pretty big. Individuals here usually require a lot of support. They might find it really hard to kick off conversations and often only respond well to very structured and clear. Their repetitive behaviours are often very marked and can really interfere with how they perform everyday activities.

These severity levels help clinicians understand how much someone is struggling. They can then adjust the support and help that person's needs.

To find out if someone has autism, we mostly use a few tools: the Autism Diagnostic Observation Schedule (ADOS), Autism Diagnostic Interview-Revised (ADI-R), and the Childhood Autism Rating Scale (CARS). These tools help doctors check and diagnose autism based on what they see and how people respond during interviews.

In the ADOS assessment, which is partly structured, doctors watch how someone behaves in different social situations. This part is important! It helps them see how well a person interacts with others, how they play, any repeated behaviours, and how they communicate.

The ADI-R is a detailed parent interview. It is semi-structured, which means it has a clear guide but allows for some flexibility. This interview gathers information about a person's early development, social skills, and what they do now. It is really helpful! You can learn a lot about someone's growth and how they are right now.

Now, let us talk about the CARS. This is a rating scale by a clinician. They use comments from caregivers and direct observations to complete it. The CARS looks at much behaviour related to ASD. It checks how someone interacts socially, their communication skills, and even repeating actions or behaviours.

However, these assessment tools are typically owned by the company and may come with a price tag as an added challenge. The use of ADOS-G and ADI-R usually necessitates training and official authorization to guarantee consistent administration and understanding [8, 9]. The accessibility and availability of these diagnostic assessments can vary based on factors such as geographical location and available resources; despite their associated costs and certification prerequisites, they continue to be essential aids in diagnosing ASD.

The creation of instruments such as the INCLEN Diagnostic Tool for ASD and the Indian Scale for Assessment of Autism (ISAA) [10, 11] marks a notable advancement in meeting the unique requirements and circumstances of diagnosing autism in India. These tools are customized for the populace and cultural setting to tackle constraints linked to globally accessible tools.

By offering these tools at no cost, to the public in India it helps reduce obstacles to diagnosing and providing support for individuals with autism. This increased access can result in detection and intervention which plays a role in enhancing outcomes for children diagnosed with autism.

Furthermore, the use of these tools for standardizing diagnosis can play a role in maintaining consistency and precision during the assessment process. This can ultimately result in dependable diagnoses and suitable interventions being implemented.

In general the creation and accessibility of these tools play a role in enhancing autism diagnosis and treatment in India, ultimately improving the quality of life for people with autism and their families.

6.3.5 Screening of ASD

There is a comprehensive list of internationally standardized and validated autism screening tools that are also used in India. These screening tools play a vital role in the early identification of children at risk of ASD [12]. Early identification allows for timely intervention such as sensory stimulation, behavioural therapy, and educational support, which can significantly improve the outcome of children with ASD.

Commonly used screening tools are:

1. *Modified Checklist for Autism in Toddlers (M-CHAT):* This is a widely used screening tool for toddlers aged between 16 and 30 months. It consists of a parent-reported questionnaire designed to identify early signs of ASD.
2. *Autism Spectrum Quotient (AQ):* The AQ is a self-administered questionnaire for adults designed to measure traits associated with ASD. It assesses various aspects such as social skills, communication, attention to detail, imagination, etc.
3. *Communication Skills Questionnaire (SCQ):* The SCQ is a parent-reported questionnaire used to screen for autism in children as young as 4 years of age and used to assess speech and communication skills, social skills, and repetitive behaviours.
4. *Social Responsiveness Questionnaire (SRS):* It is a questionnaire used by parents, teachers, and/or caregivers for social and behavioural problems found in ASD at all ages.
5. *Autism Behaviour Checklist (ABC):* The ABC is a checklist filled out by a parent and/or caregiver to assess various behaviours linked to ASD including social interactions, communication skills, and oppositional behaviours.
6. *Social Communication Disorder Checklist (SCDC):* This checklist is used to screen children for communication difficulties. It assesses functional language skills, social interaction, and non-verbal communication.

These screening instruments/tools support educators and medical professionals in identifying children who might benefit from additional ASD assessment and intervention. Helping children with autism requires early screening, identification, and intervention. These steps improve the quality of life for autistic people and their families and lead to better outcomes. Individualized therapy and educational programmes can assist in addressing issues and enhancing strengths.

6.4 Occupational Therapy and Autism Spectrum Disorders

Autism spectrum disorders (ASD) presents with a diverse range of symptoms and signs. Some individuals may develop typical speech and language but face challenges in social skills and behaviours, while others found difficulties in self-control, sensory processing, etc. Besides ASD is a spectrum disorder, which means every person who has it has a different set of requirements, strengths, and weaknesses. For instance, some may excel in specific areas like math, music, or art and at the same time might struggle in various aspects of life, including social skills, daily living activities, and vocal control.

One of the most significant impediments associated with autism spectrum disorder is the inability to comprehend sociability and daily tasks not only social but also daily tasks; then the emotional, mental, and behavioural struggles; the repetitive behaviour; and, on some days, various issues.

There are a few individuals who go through specific disorders apart from the hyperactivity such as ADHD. They may be suffering from anxiety, depression, and stress. To overcome these diseases, a number of strategies (individual and family member-centred treatment) have been recommended by medical professionals.

Yet with the help of the right treatment and individualized recovery approaches, many kids with autism can reach better results. Through earlier treatments include behavioural therapy, speech therapy, occupational therapy, and education support, people with autism may be able to develop their social, emotional, and cognitive skills.

It is very critical to identify and make use of the positive sides and capabilities of the people with autism and provide them with the help, aid, and printable tools they need to succeed in their lives. People with mental health disorders can be positive individuals through understanding, tolerance, and having access to the right treatment.

The use of a multidisciplinary team approach (MDT) for the management of obsessive-compulsive disorder (OCD) associated with autism spectrum disorder (ASD) is widely encouraged. In this way, a variety of professionals cooperate to meet the individual needs of the children with ASD.

Methodological research is an integral pedagogical activity. An occupational therapy practitioner is mainly concerned in assisting children with autism in learning how to engage in meaningful activities and cope in their surroundings. This consists of good hearing, personal hygiene, motor coordination, and basic living skills.

Other professionals that are involved in an ASD MDT may include:

1. *Behaviour therapist:* The behaviour therapist uses applied behavioural analysis (ABA) techniques to teach and reinforce behavioural skills and knowledge, reduce behaviour problems, and strengthen relationships and communication.
2. *Speech-language therapists:* Improves communication skills—language (concepts and understanding), talking (expressing, verbal fluency), non-talking (body language), and speech sounds.

3. *Psychologist or psychiatrist:* Can help with psychological assessment and some people with Down syndrome have mental health issues like depression or anxiety so they can get support there and also advice on managing behaviours.
4. *Special educators:* They provide educational assistance specially designed to meet the unique needs of students with autism under an individualized education programme (IEP).
5. *Paediatrician or developmental paediatrician:* They provide medical treatment, monitor growth and development, and coordinate medical care among multiple specialists for patients.

In the MDT model of service delivery, the knowledge and skills of several disciplines are combined to provide solutions for children with ASD. It is based on the principle that each child diagnosed with autism has his or her own unique strengths and areas for self-help and inclusion; therefore, effective service must differ by individual.

6.4.1 Role of Occupational Therapy

The OT professionals play an important role in the design and implementation of interventions for people with ASD and their families. Children with ASD often face challenges that impact their ability to participate fully in a variety of activities at home, school, and in the community. Occupational therapy were trained to address these challenges and assist persons with ASD and develop the skills they need to participate in some meaningful activities and progress their life's quality.

Occupational performance issues that occupational therapy commonly address when working with individuals with ASD include:

1. *Sensory processing issues:* Many children with ASD experience sensory hypersensitivity or difficulty processing sensory information, such as sensitivity to sounds, textures, or lights. Occupational therapy can provide sensory integration therapy and sensory stimulation interventions to help individuals regulate their sensory responses and better participate in daily activities.
2. *Development of fine and gross motor skills*: Some children with autism may have delays or difficulties developing fine motor skills (e.g. handwriting, handling small objects) or gross motor skills (e.g. coordination, balance). Occupational therapy can provide activities and exercises to improve motor coordination, strength, and dexterity.
3. *Self-care skills*: Individuals with autism may need support developing self-care skills, like dressing, grooming, and feeding. Occupational therapy can teach adaptive strategies and provide structured routines to help persons with ASD become more independent in these activities of everyday living.
4. *Social skills and peer interaction*: Occupational therapy can facilitate social skill training and provide opportunities for individuals with ASD to practice social interactions with communication skills in a structured setting. Role-playing,

group play, and community stories can be used to teach and encourage good behaviour.

5. *Functional and environmental modifications:* Therapists may recommend modifications to the home, school, or community setting to support the individual's participation and independence. This might involve providing assistive devices and adaptive tools or creating an environment tailored to individual needs.

Occupational therapy support individuals with ASD through personalized interventions aimed at work and employment. These tailored approaches help them build the skills and confidence necessary to engage in daily activities and reach their full potential.

6.4.2 Role Priority and Team-Based Approach

OTs as team members are benchmarking professionals (occupational therapy) whose main purpose is to uplift these children suffering mental disorders all through their lifespan. They concentrate on the individualization of the programme, including the persons with autism, and maximizing their autonomy and quality of life.

Therapeutic programmes for children with autism can be developed and implemented by OT therapists in their leadership roles. They may also work with other specialists, such as a speech therapist, behaviour analyst, and teachers, to provide support.

In professional support, OT therapists work directly with individuals with ASD and their families to develop a range of skills and abilities, including cognitive and motor skills, self-care, and relationship and behaviour modification. Children with disabilities often benefit from a variety of interventions, such as hearing aids, behavioural interventions, and technology, to support their growth and abilities in daily life.

In general, occupational therapy plays an important role in supporting people with autism in many ways, helping them develop the skills and ideas they need to participate in society and live successful lives.

Various interventions and training should be provided to highlight the important role of professionals in meeting the special needs of children and youth with autism. In fact, occupational therapy play an important role in supporting people with ASD problems at different developmental stages and focusing on various tasks.

Occupational therapy for children with autism usually focus on interventions to improve hearing, social and cognitive, and physical work, personal care, and play participation. These interventions may include cognitive integration, community education, play interventions, and activities to promote independence in daily living.

As children with ASD enter adolescence, continuing education goals may focus on behavioural and social skills, preparation for transition to employment or higher education, and promoting social independence. Occupational therapy will address functional skills, coordination, and adaptive strategies to support the transition to adulthood.

In many of these sessions, Occupational therapy work closely with other professionals as part of a collaborative team. This team approach provides comprehensive assessment, individual goal setting, and comprehensive planning to meet the needs of children with disabilities. Collaboration with professionals may include behavioural therapists, coaches, psychologists, speech therapists, and physicians.

By integrating occupational therapy into an integrated treatment plan, children and youth with mental health problems can receive effective support tailored to their individual strengths, challenges, and growth trajectories. These integrated approach results in better interventions and helps in best outcomes for people with ASD across different stages of their lives.

6.4.3 Occupational Therapy Objectives

The primary objective of OT for an individual with ASD is to influence interactions in activities and make the individual more independent in settings like family, school, and community life. Occupational therapy use an integrated, client-centred approach to find effective ways of dealing with the needs and problems of people with autism. This happens in such a way that their individual interests and strengths are enhanced, while interests, aptitudes, and environmental factors are taken into consideration.

Occupational therapy cooperate with children with autism and their families to find the most important and interesting activities that fit them in their daily lives. In the process of accomplishing the activities, people with ASD enhance and develop their skills, confidence, and independence, circulate among others, and heighten the quality of life.

The treatments may incorporate different methods such as social work, behavioural interventions, cognitive-behavioural therapy, cognitive integration and behavioural therapy behaviour, and adaptive equipment or technology. Professionals work with other professionals, the family, and the broader society to avail a very comprehensive support system for people with autism and therefore help them to improve their life.

Occupational therapy enables individuals with ASD to attain a problem-solving, participation, and personal capability lifestyle and hence live productively in society. The curative method encompasses cultures and individuals, stressing the importance of self-care and healthcare improvement as a determinant of life quality.

6.5 Application of Occupational Therapy Interventions

6.5.1 Gathering Data

Standardized assessments, questionnaires, and expert opinions are essential tools used with the help of Occupational therapy to guide autistic adolescents and their families in the development and use of intervention. These assessments provide valuable information about a person's strengths, challenges, preferences, and needs, permitting clinicians to advance intervention plans that focus on specific issues and promote true practical knowledge.

These assessments help identify areas of strength and areas of need for assistance and create a personalized treatment plan based on the unique needs of individuals with ASD. A personalized treatment plan is created for each autistic person based on their specific needs. Look at it from different angles.

These exams provide a better understanding of a person's work skills, their problems, and their techniques for solving everyday tasks. These interventions may also include integration, physical education, social development, personal education, environmental modification, and techniques that promote independence and participation in everyday life.

Interventions should be effective and relevant to meet the wishes and dremas of individuals, their familiees, OTs' and other professionals. Autistic people and their families face challenges in everyday life.

Through evidence-based programmes and interventions, therapists work to assist individuals with autism realize their potential, enhance their quality of life, end up more independent, and participate in life.

6.5.2 Occupational Performance Areas Addressed

Occupational therapy professionals play a vital role in enhancing the functioning and well-being of individuals with autism spectrum disorder (ASD). They work in many areas to help people with autism achieve more independence, improve their quality of life, and enhance their overall health.

1. *Activities of daily living (ADL):* Occupational therapy monitor and assist adolescents with self-care tasks such as bathing, dressing, showering, and feeding. They may also provide adaptive equipment, teach independent skills, or adjust the environment to facilitate self-care and daily living.
2. *Instrumental activities of daily living (IADL):* Practitioners work to improve basic skills for independent living, such as meal preparation, housework, financial management, and transportation. Occupational therapy help people increase these skills to improve their potential to live independently and take part in community life.
3. *Adaptive behaviours:* OT professionals focus on developing adaptive behaviours important for daily functioning, including communication, social skills, self-regulation, problem-solving, and stress management. Through mediation, they encourage the development of these skills to help people with ASD cope with daily challenges.
4. *Rest and sleep:* Occupational therapy assess sleep patterns, behaviours, and environments that may impact quality sleep and provide effective strategies and interventions to promote better sleep and improve overall sleep and health for people and their families.

5. *Employment/pre-employment development:* For youth and adults with mental illness, Occupational therapy assist with career development and employment planning, including job evaluation, educational career advice, skill training, resume writing, job search strategies, and interview preparation. Their goal is to increase productivity and performance.
6. *Social participation*: Therapists work to enhance social engagement and participation in meaningful activities, offering social skill training, opportunities for peer interactions, and strategies for community integration to support better social relationships and active participation in social environments.

By addressing these critical areas, occupational therapy help individuals with ASD achieve greater independence, improved quality of life, and enhanced overall well-being. Through client-centred and evidence-based interventions, they support individuals in reaching their goals and participating more fully in their communities.

6.5.3 Occupational Performance Components and Skills Addressed

Furthermore, occupational therapy work on improving various foundational skills that underlie the challenges faced by children with ASD [13]. Here is how Occupational therapy approach these key skills:

1. *Gross motor skills:* Occupational therapy focus on developing gross motor skills, such as coordination, balance, strength, and resistance. Activities can also include exercise, games, sports, and outdoor games to improve normal body function and mobility.
2. *Fine motor skills:* Fine motor skills involve the coordination of small muscle tissues in the hand and fingers to perform complete tasks such as writing, drawing, cutting, and manipulating objects. Occupational therapy use crafts such as play dough, puzzles, drawings, and crafts to improve motor control, dexterity, and accuracy.
3. *Visual-motor integration (VMI) skills:* VMI refers to the ability to coordinate visual capture with motor skills, such as hand-eye coordination and copying shapes or letters. Occupational therapy appreciate doing things like spotting, tracing, colouring, and printing to improve body-hand-eye coordination.
4. *Visual awareness:* Visual awareness involves the ability to interpret and understand visual information, such as determining relationships between shapes, patterns, and spaces. Therapists work to improve vision through activities that use visual perception, discrimination, visual perception, and visual memory.
5. *Sensory processing:* Sensory processing is where the brain receives, interprets, and responds to information from the environment. Therapists can help people manage stress through a combination of sensory processing, mindfulness, and attention interventions.

6. *Behavioural therapy:* Therapists work to encourage self-regulation and emotional control so that people with mental illness can control their behaviour, thoughts, and reactions to events. This may include teaching coping strategies, relaxation techniques, kindness, and relationships to improve behaviour.

Occupational therapy can help individuals with autism spectrum disorders improve their overall functioning, independence, and participation in daily activities by addressing the following issues and developing wisdom. These interventions are individualized, tested, and tailored to the unique needs and goals of each individual with ASD.

6.6 Approaches Used by Occupational therapy in Autism Spectrum Disorders

OT plays an important role in helping individuals with autism develop skills and strategies to improve daily functioning and well-being. Clinicians use the following treatments for individuals with autism:

1. *Sensory integration therapy:* Many people with autism spectrum disorder have hearing loss or difficulty processing auditory information. Sensory integration treatment is used by Occupational therapy to improve persons' comfort in daily activities by regulating their responses to sensory stimulation.
2. *Improve fine motor skills and gross motor skills:* Some people with autism spectrum disorder may have problems with fine motor skills (such as writing, buttoning a shirt) or driving skills (such as coordination, balance). Occupational therapy provide interventions to improve motor skills, including exercise, activities, and therapeutic strategies.
3. *Social skill training*: Occupational therapy work to develop social and language skills to help people with autism spectrum disorders manage relationships. This includes games, community stories, and group activities to improve communication and relationships.
4. *Daily living education (ADLs) and self-care training:* Occupational therapy help people with autism independently perform daily living tasks such as dressing, exercising, bathing, and eating. They may also provide replacement equipment or teach other ways to improve personal hygiene.
5. *Behavioural strategies:* Occupational therapy work with individuals with ASD, their families, and other professionals to develop behavioural strategies and interventions that help manage negative behaviours and promote positive behaviour change.
6. *Cognitive-behavioural strategies:* Occupational therapy may also consist of cognitive-behavioural techniques to manage stress, anxiety, and problem-solving in people with autism.
7. *Environmental modification:* Professionals examine the person's home, school, and community environment to discover limitations to their participation and

suggest changes or services they need. It may also include creating an emotional environment or using visual aids.

8. *Play in occupational therapy:* As an occupational therapy for mental disorders, play is designed to develop a variety of skills and improve interpersonal relationships, communication, thinking, and physical coordination.
9. *Task-oriented treatment approaches:* Task-oriented treatment approaches for autism focus on helping children develop the skills they need to complete specific tasks. These approaches are designed to improve functional abilities and enhance independence in daily activities. By breaking down complex tasks into smaller, manageable steps, occupational therapy can help children with autism develop the necessary skills to complete tasks successfully.
10. *Transition planning and vocational rehabilitation:* Occupational therapy assist adolescents and young adults with ASD with transition planning for postsecondary education, employment, and unbiased living. They assist develop vocational skills, community integration strategies, and support networks.
11. *Visual supports and strategies:* Visual supports like schedules, visual cues, and picture-based communication systems (e.g., PECS [picture exchange communication system]) are commonly used by OTs to enhance communication, increase understanding of routines, and promote independence.
12. *Family education and support:* Occupational therapy provide education and support to families, caregivers, and educators on understanding ASD, implementing effective strategies, and advocating for the individual's needs.
13. *Collaboration with other professionals:* OTs collaborate with other professionals involved in individual care, such as speech therapists, behaviour therapists, educators, and medical professionals, to ensure a holistic method of treatment.

Overall, occupational therapy interventions are individualized, evidence-based, and attentive to improving the functional abilities and quality of life of persons with ASD across the lifespan. By addressing sensory, motor, social, and behavioural challenges, occupational therapy help persons with autism spectrum disorder achieve greater independence, participation, and overall well-being.

Some of the more commonly used approaches used by occupational therapy are described in the following sections.

6.6.1 Sensory Integration and Sensory-Based Interventions

Occupational therapy often use sensory integration therapy and sensory-based interventions to address sensory processing difficulties in persons with ASD. Sensory processing difficulties are highly prevalent in ASD with up to 96% of children affected [14–16]. These difficulties are now recognized in DSM-5 [2, 17]. The majority of individuals with ASD experience hypo- and hyper-reactivity to sensory inputs, impacting their daily routines. Current estimates suggest that over 80% of individuals with ASD exhibit accompanying sensory processing issues. This underscores the significant impact of sensory processing challenges on persons with ASD and highlights the importance of addressing these difficulties in interventions and support strategies.

Occupational therpay practitioner Jean Ayres formulated the sensory integration theory [18], drawing from neuroscience, developmental psychology, education sciences, and occupational therapy. This theory underpins sensory integration therapy, widely used in paediatric occupational therapy, particularly with persons diagnosed with autism. A survey of occupational therpay practitioner revealed that 99% referred to sensory integration therapy for such cases [19]. The theory posits four key principles:

1. *Sensory-motor development is crucial for learning*: Sensory-motor skills, such as fine motor, gross motor, and sensory integration, lay the foundation for learning and cognitive development. Children with autism often experience challenges in these areas, which can impact their ability to learn and participate in daily activities.
2. *Interaction with environments shapes brain development*: The brain develops and adapts in response to interactions with the environment. This concept is called neuroplasticity. Occupational therapy can help children with autism overcome challenges and reach their potential by creating environments and activities that support brain development and change.
3. *The neurological system exhibits plasticity:* Neuroplasticity is the brain's ability to adapt to new experiences, learning, and changes in the environment. This concept is important for children with intellectual disabilities because it shows that the brain can be rewired and compensated for through intervention and treatment.
4. *Meaningful sensory-motor activities strongly mediate plasticity:* Meaningful sensory-motor activities, such as play, exploration, and socialization, promote neuroplasticity and support brain development. Occupational therapy can tailor activities to each child's needs and interests and use the power of sensorimotor skills to promote positive and adaptive change.

Children develop the skills and abilities they need for academic success, social interaction, and daily functioning. SI therapy is designed to give children the control and flexibility to respond independently and appropriately and cope with these emotions.

Sensory integration, according to this hypothesis [20], is the process of organizing sensory information in the brain to generate adaptive responses. SI treatment attempts to offer children with controlled and meaningful experiences so that they can respond spontaneously and properly and effectively integrate the sensations [21]. These principles are supported by evidence from OT practice and research that suggest the importance of sensory integration in facilitating developmental and learning processes in children, particularly those with autism.

6.6.1.1 Overview of Sensory Integration Therapy and Sensory-Based Approaches

Sensory integration therapy, developed by A. Jean Ayres, is based on the concept that the brain can adapt and organize sensory information for more efficient processing. The therapy aims to assist persons with ASD to better process and respond to sensory stimuli from their environment. It typically involves engaging individuals in structured sensory activities designed to provide sensory stimulation in a controlled and therapeutic way.

6.6.1.2 Evidence Supporting Sensory Integration Therapy

1. *Systematic reviews and meta-analyses*
   - Several systematic reviews and meta-analyses have examined the effectiveness of sensory integration therapy for persons with autism spectrum disorder. While findings have been mixed, some reviews suggest positive effects on sensory processing, adaptive behaviours, and motor skills [22, 23].
2. *Randomized controlled trials (RCTs)*
   - Several RCTs have reported positive outcomes of sensory integration therapy for children with ASD. For example, the study by Pfeiffer et al. (2011) found improvements in sensory processing and social participation following sensory integration intervention compared to a control group [24].
3. *Parent and clinician reports*
   - Many parents and clinicians report subjective improvements in sensory processing difficulties, behaviour regulation, and participation in daily activities following sensory integration therapy.

While evidence supporting sensory integration therapy for persons with ASD exists, it is essential to acknowledge that the quality of research studies varies, and more high-quality research is needed to further validate its efficacy. Additionally, sensory-based interventions encompass a broad range of approaches beyond sensory integration therapy, including environmental modifications, sensory diets, and specific sensory activities tailored to individual needs.

6.6.1.3 Applications of Sensory Integration and Sensory-Based Interventions

The National Institute of Child Health and Human Development (NICHD) in 2005 has conducted research indicating that individuals with ASD often process sensory information in different ways than naturally developing children. This includes how we see and respond to the various senses, such as touch, sound, smell, and taste.

Hearing differences in children with autism can manifest in a variety of ways, including increased sensitivity to certain stimuli or difficulty filtering content. These variations can have an effect on everyday things to do and lead to interpersonal, communication, and behavioural problems.

Understanding the variations in needs is essential to develop positive support strategies and interventions for people with autism. Occupational therapy and other experts who work with people on the autism spectrum play an important role in assessing the child's needs and providing interventions based on the child's hobbies and ideas.

The mission aims to expand support for adolescents with autism by thinking about different perspectives. A conversational approach focuses on evaluating and fitting hearing aids to meet the needs of each child.

Using techniques such as paying attention to noise or planning ahead to avoid getting dressed can improve the child's daily routine. These small changes can make a big difference in their treatment and their ability to participate in activities.

Occupational therapy play a vital role in helping families navigate the challenges of raising a child with autism. Occupational therapy can have a significant impact on the lives of family members by adapting activities and environments to the child's needs and working with families to engage them in their daily lives.

Self-care tasks such as eating, bathing, grooming, and sleeping can be particularly difficult for youth with ASD and their families. Occupational therapy may provide valuable assistance in the following ways:

1. *Environmental adaptation:* Modify the home surroundings to limit stress and create an environment that promotes calm.
2. *Learn new skills:* Break self-care tasks into manageable steps and use visual aids to assist children recognize and comply with the process.
3. *Perform calming or awakening activities:* Introduce activities that assist you control your emotions, such as questioning or reflection.
4. *Work with the family:* Work with parents and caregivers to develop techniques that respond to the needs and interests of the child.
5. *Provide education and support:* Provide families with ASD information and techniques to help their child develop and live in everyday life.

Occupational therapy can assist children with autism and their families deal with necessary aspects of daily life to limit stress and thus improve physical condition and well-being in a healthy way.

In schools, therapists play a vital role in helping students who face a variety of challenges, including emotional, behavioural, developmental, and learning disorders. Working with teachers and other school staff, therapists can help adjust classroom activities and environments to help students participate in style and succeed in learning.

Here are some approaches that professionals can take to help students in school:

1. *Adapt the activities and the environment:* Professionals (occupational therapy) can modify the activities and substances used in the study room to meet the needs, physical abilities, and emotions of the students. This may also consist of seating options, appropriate equipment, or visual aids to help students stay active and engaged.
2. *Educating teachers and faculty:* Occupational therapy can propose residents and instructors about listening to issues that have an effect on students' everyday work. They can also help limit disruptive behaviour and create more supportive

learning experiences by emphasizing and providing classroom management strategies.

3. *Collaborate to strengthen an individualized education programme:* Occupational therapy can work with the school's specific learning team to develop an individualized education programme (IEP) or Model 504 for students with disabilities or developmental disabilities. This consists of dreams and plans to help college students academically, socially, and emotionally.
4. *Provide direct intervention:* Sometimes occupational therapy provide direct intervention to students through individual or group therapy. These instructions focus on enhancing coordination, motor coordination, self-regulation, or skills associated to everyday activities.

Therapists can also work with teachers, teaching assistants, and other faculty team members to assist create collaborative and supportive learning surroundings in which everyone can thrive. Many persons with autism spectrum disorder (ASD) thrive in environments filled with positive, hands-on activities.

A supportive surroundings and structure can go a lengthy way toward enhancing the excellent of lifestyles for these children. Dalrymple's 1995 review shows how a supportive environment can help autistic children by helping them cope with the changes of daily life in addition to time constraints.

Four types of environmental support have been identified that aim to differentiate the environment of autistic children and enhance their first degree of life: temporal support, technical support, spatial support, and mental support:

1. *Temporary support:* These supports are usually temporary and give children time to prepare things to do or changes in their everyday life. This may also include visual, physical, or verbal cues to sign modifications and future events, supporting teens adapt to their needs and experiences.
2. *Procedural support:* Procedural support entails explaining behaviours and supporting children with autism recognize what to anticipate in distinct situations. Providing clear and structured instructions, breaking down the task into manageable steps, and using sentences or descriptions can help facilitate perception and engagement in the task.
3. *Spatial support:* Spatial support involves modifying the physical surroundings to facilitate crawling and walking for adolescents with autism. It may also include creating designated areas for specific activities, using visual cues to mark boundaries or paths, and minimizing distractions to minimize distractions.
4. *Assertion support:* Assertion support services are designed to assist children with autism take part in society and express their needs and interests. This may also include instructing social skills, writing, or encouraging conversation to initiate and inspire conversation and build peer relationships where necessary.

By integrating a supportive surrounding into the daily life and activities of young people with autism, caregivers, teachers, and clinicians can create a predictable, structured, and supportive environment that improves the child's quality of life and general health.

Different types of environmental support can benefit children with autism:

1. *Visual schedule:* Providing visual time for the duration of activities to do or tasks can assist children with autism recognize and predict what will show up next. A visual agenda provides a clear structure to the day, reduces stress, and will increase motivation.
2. *Consistent routines:* Establishing steady routines, transitions, and study room routines can assist persons with autism experience secure and in control. Relationships furnish predictability, which can decrease stress and enhance behaviour.
3. *Structured activities:* Structured activities with clear goals and expectations can assist young people with autism remain targeted and engaged. Providing step-by-step directions and breaking tasks down into chunks can enhance perception and participation.
4. *Visual supports:* Using visual supports such as visible time, visible cues, and visible cues can enhance conversation and perception in children with autism. Visual supports furnish vital facts that will assist make clear expectations in quite a number of areas.
5. *Sensory-friendly environment:* Create emotional surroundings that minimize frustration and promote remedy and happiness via meeting the wants of children with ASD. This may also encompass presenting listening to aids or a quiet managed environment, as properly as light and sound controls.

Collaboration with parents and instructors are integral to growing surroundings that helps children's needs. Occupational therapy can work with adolescents to facilitate social interplay throughout the surroundings and to assist them take part in everyday activities.

6.6.2 Cognitive and Gross Motor Skill Development

Occupational therapy (OT) performs a vital function in helping the development of cognitive and gross motor skills in children with autism spectrum disorder (ASD). Practitioners, with the use of evidence-based strategies, can focus point on the following areas:

6.6.2.1 Developing Fine Motor Skills

- *Handwriting Skills:* Occupational therapy frequently work with children with autism to strengthen writing skills, which can be difficult due to handholding, and driving arrangements. Interventions consist of workouts to fortify muscles, enhance hand-eye coordination, and exercise splendid writing techniques.
- *Manipulative Skills:* Additional getting to know focuses on developing skills to use small objects, such as scissors, string, or hair clips. These things, done with

assistance, enhance the motor coordination and flexibility needed for independence in everyday life.

- *Cognitive Integration:* Children with autism spectrum disease (ASD) regularly have cognitive deficits that have an effect on quality motor skills. Occupational therapy use a range of techniques to tackle emotional issues and enhance planning and motor control.

6.6.2.2 Development of Gross Motor Skills

- *Balance and Coordination:* Occupational therapy use movement to enhance balance, coordination, and physical awareness in adolescents with autism. This consists of exercise, sports, and physical activities to do such as jumping, hopping, and climbing.
- *Core Strength:* Core energy is necessary for retaining stability and energy at some stage in exercise. Extremity interventions may additionally encompass muscle strengthening workouts such as yoga poses, Pilates, and taking part in with therapy equipment.
- *Motor Coordination:* Children with autism spectrum ailment may additionally have motor coordination troubles that have an effect on capital movements. Occupational therapy can use these activities and visual aids to assist students learn and function effectively.

Overall, occupational therapy services for children with ASD that are precise to their needs can assist in enhancing abilities and regular abilities, thereby enhancing work and quality of life.

6.6.3 Social Learning Skill Training

Social skill training is a necessary intervention for individuals with autism. Children with autism regularly have poor language capabilities and have problems with perception of non-verbal cues such as eye contact and facial expressions. These troubles can lead to isolation, bullying, and bad relationships [25]. Therapists play a necessary role in fixing the social issues of children with autism with evidence-based interventions.

The foremost aim of cognitive skills is emotional intelligence, which means grasping and deciphering the thoughts, feelings, and information of others [26, 27]. Cognitive impairment is common in children with autism and affects understanding of emotions and feelings. Occupational therapy use a variety of strategies to target mental illness, including interventions, community stories, and activities.

Executive functioning [28, 29], another neuropsychological model, is also implicated in social skill deficits in autism. Executive functions such as self-regulation, planning, and flexibility play a crucial role in social interactions and adaptive behaviour. Occupational therapy work to enhance executive functioning skills through structured activities and interventions tailored to the individual requirements of children with ASD.

The weak central coherence model advocates that persons with ASD have difficultly integrating info from various sources, leading to challenges in processing and understanding social cues [30, 31]. Occupational therapy address weak central coherence by providing structured interventions and support to assist children with autism make sense of social information and navigate social situations more effectively [32, 33].

Social skill training interventions implemented by occupational therapy target a range of skills, including basic social interactions, communication skills, friendship skills, and emotion recognition and management [34–37]. These interventions may involve individual or group sessions, parent training, and collaborative efforts with teachers and caregivers. Evidence-based techniques, for instance, social stories, video modelling, and peer mediation, are usually utilized in occupational therapy interventions to advance social skills in children with ASD [38–42].

Assessment of social skills in children with autism involves standardized tests, observations, and interviews with parents and caregivers [43]. Occupational therapy conduct comprehensive assessments to identify strengths and areas of social functioning that need improvement in social functioning. An intervention plan is then developed based on the individual's requirements and goals, focusing on a client-centred, evidence-based approach that promotes active engagement and participation in social activities.

The goal of these interventions is to improve relationships, communication, and behaviour, thereby increasing the child's ability to socialize and form relationships. Some of the techniques doctors use to improve the socialization of children with intellectual disabilities include:

1. *Structured socialization programmes*: Doctors often use social services designed to meet the needs of children with autism. These projects often include hands-on learning, simulations, role-playing, and creative writing to teach specific skills and ideas.
2. *Social stories:* Community stories are narratives that provide examples of interpersonal relationships and appropriate responses to individuals with autism. Practitioners can use community testimonies to carry leadership, hope, and emotional awareness.
3. *Pre-mediated interventions:* Occupational therapy can enhance the social and emotional well-being of children with autism by encouraging effective pre-mediated interventions. These interventions consist of socializing children with autism, providing everyday support, and providing social activities.
4. *Video modelling:* Video modelling entails watching and playing videos that educate ethical behaviour. Occupational therapy can use video modelling as a tool to teach children social skills and promote social learning.

   Occupational therapy frequently use a range of evidence-based strategies to furnish social education to children with autism. These interventions are designed for every child's special social, communication, and energy improvement needs.

   In conclusion, social studying is a vital section of the therapy format for people with autism. Occupational therapy assist children with autism enhance the relationships they want to be successful and thrive through addressing negative emotions, difficulties, and dysfunctions. Therapists play a necessary role in enhancing the social lives of persons with autism through evidence-based interventions and self-esteem.

6.6.4 Activities of Daily Living and Self-Care Skills

6.6.4.1 Brief Introduction

Activities of daily living (ADLs) include many of the basic activities that most people do every day to care for themselves and their health. These activities are the foundation for independent living and functioning in society. ADLs include bathing, dressing, eating, feeding (assisted), bladder and bowel control, physical activity, personal care, body care and hygiene, gender control, and toilet hygiene. Participation in these activities not only makes life easier, but also improves health and community participation. Mastery of activities of daily living (ADL) empowers children and youth in all areas of life. Learning these skills in a positive environment can help improve participation in school, play, social work, and ultimately work. Acquiring ADLs and IADLs enhances a person's abilities and independence, encouraging independence, confidence, and community participation. Understanding social norms and expectations allows individuals to confidently step into society. Finally, acquiring these skills allows people to contribute to their families, communities, and society at large [44].

Activities of daily living (ADLs) and essential activities of daily living (IADLs) are essential for individuals with autism to function independently and in daily life. However, people with autism often have difficulty participating in these activities, which can affect their quality of life [45–49]. Physicians have an important role in evaluating and resolving these problems with evidence-based interventions.

6.6.4.2 Assessment

Assessment of ADL and IADL performance begins with a functional assessment of the task. This will include interviews with family members, caregivers, and people with autism [50] as well as observations in a variety of settings. Assessment and rating scales, the Vineland Adaptive Behavior Scale [51] and the Paediatric Disability Rating Scale [52], can also be used to gather information about the individual's function and needs. Environmental and cognitive issues are also included in the assessment for better understanding of personal skills and difficulties.

6.6.4.3 Intervention

In order to improve ADL and IADL for persons with ASD, occupational therapy use different methods [50]:

1. *Development and maintaining performance*
   - *Behavioural approaches:* Behaviours—reinforcement and signing, that is, which are used to develop and maintain the abilities of individuals with

autism. These technologies may involve providing visual, verbal, tactile, or environmental cues for learning and completing the task.
   - *Chaining:* The use of backward chaining, forward chaining, or whole-task training in teaching through the use of techniques involves breaking down complex skills into smaller parts.
   - *Time-delay procedures:* It involves the gradual reduction in the amount of help provided and is used to foster independent task performance.
2. *Activity adaptations or compensatory methods*
   - *Grading:* The subtasks are adjusted to the level of difficulty for the child, and modifications are made to tasks based on functioning level [50].
   - *Visual strategies:* Schedules, scripts, and task analysis, in other words visual supports that are commonly implemented to aid in understanding and ordering activities. Particularly useful for individuals with autism who happen to process information more successfully when it is visual [53–56].
   - *Assistive technology:* It is the use of technological aids in realizing improved learning and skill development for interventions through the combination of virtual reality environments, video modelling, and prompting systems [56–59].
   - *Sensory techniques:* Problems related to sensory modulation and its impact on the performance of activities of daily living and instrumental activities of daily living need a sensory-friendly environment with adaptive sensory strategies in action [60].

6.6.5 Behavioural Strategies

Behavioural therapy for ASD uses the principles of behaviour intervention and positive outcomes. A key behaviour treatment for autism spectrum disorder (ASD) is applied behaviour analysis (ABA) which can be used in many settings and formats, schools, clinics, and family programmes. Some specific types of ABA are struggling training [61] and Lovaas therapy [62], both of which are a practice-based approach.

This can happen if the child is already in an ABA programme or if the child is learning a specific skill that would benefit from ABA methods such as toilet training or shoe tying.

6.6.5.1 Applied Behaviour Analysis (ABA)

- ABA is a scientific method that focuses on analysing the antecedents, actions, and consequences of behaviour and using that analysis to modify behaviour.
- Reinforcement, shaping, prompting, and fading are ABA techniques that help teach new abilities and lessen problematic behaviours.
- ABA programmes are individualized primarily based on the individual's needs and strengths, and progress is monitored systematically through data collection and analysis.

6.6.5.2 Discrete Trial Training (DTT)

- DTT is a structured teaching approach used in ABA that breaks down skills into small, manageable steps and teaches each step systematically.
- DTT is giving a child a series of trials, each of which consists of a cue or explicit instruction, a consequence (e.g. reinforcement or correction), and the child's response.
- DTT is often used to educate foundational skills, such as language, social, and self-help skills.

6.6.5.3 Lovaas Therapy

- Lovaas therapy, also known as Early Intensive Behavioural Intervention (EIBI), is a specific form of ABA developed by Dr. Ivar Lovaas.
- Lovaas therapy typically involves intensive one-on-one teaching sessions, often conducted for 20–40 h per week.
- The goal of Lovaas therapy is to promote rapid skill acquisition and minimize problem behaviours through intensive, structured intervention.

Inclusive services may combine behavioural interventions with other therapies to support the development and well-being of individuals with ASD. Occupational and physical therapists may collaborate with ABA therapists and other professionals to manage intervention strategies that meet the diverse needs of children with autism in many workplaces.

From this perspective, the main principle of ABA is of high importance to facilitate and teach ASD-affected individuals to learn new skills, curtail bad behaviours, and enhance independence and involvement in community activities. Application of ABA principles in occupational and physical therapy sessions enhances the effectiveness of interventions based on ABA and ensures that it has a positive impact on the lives of people with autism.

6.6.6 Cognitive-Behavioural Strategies

6.6.6.1 Cognitive-Behavioural Approach Overview

The cognitive-behavioural approach posits that a person's thoughts and beliefs impact their behaviour. By making them respond to the rational thoughts by altering and modifying their irrational thoughts, it would change their behaviour [63]. This is mostly the case of the ASD patients who tend to have trouble dealing with cognitive processing and societal learning [64].

Cognitive-behavioural therapy (CBT) is a time-limited goal-oriented treatment that aims to reduce cognitive distortion, maladaptive behaviour, and emotional distress by identifying and modifying unhelpful thoughts and beliefs. Even though CBT has gained much research attention and application in many mental health disorders, its use on ASD cases has just started to gain momentum.

CBT is designed to enable the autistic person to control his or her emotions, to work on impulse control, and to tackle fears and anxiety. CBT involves changing perceptions and knowledge about behaviour [65]. Occupational therapy apply the model of CBT in their practice. These include the human occupational model and the biopsychosocial model. The occupational therpay practitioner will try to change knowledge and behaviour in diverse settings and activities [63, 66, 67].

Assessment tools of occupational therapy: Apart from regular assessments, various other special tools are used by OTs for mentally disordered people such as five-point clocks, electronic cards and video-modelling [68–70].

Occupational therapy employ a variety of programmes and activities to be met with the mentally ill. These activities may involve educating them on self-management, emotional exploration, coping with stress, and enhancing one's communication and problem-solving capacities, among others. Tools such as the Goal Attainment Scale (GAS) measure the effectiveness of the intervention by measuring the progress toward personal goals as a result of the intervention [71].

This approach is holistic in nature because it understands the individual's peculiar problems in mental illness and seeks to help him control his thoughts, behaviours, and interaction with people around him.

6.6.6.2 Targeted Areas in CBT for ASD

- Cognitive flexibility: Teaching persons with ASD to recognize and challenge rigid thinking patterns, improve problem-solving skills, and adapt to changes in routines or expectations
- Social skills: Develop social skills, intuition, and communication skills to improve interpersonal and social relationships
- Emotional regulation: Teaches relaxation techniques and coping strategies for stress, depression, and other common stressors in people with ASD

6.6.6.3 Adaptations for People with Autism Spectrum Disorder

- Visual supports: Provide visual aids such as visual aids, community stories, and visual cards to enhance understanding and encourage education.
- Concrete examples: Provide examples and clear explanations to help people with mental illness understand complex concepts and use CBT techniques effectively
- Structured format: Uses clinical and predictive models to meet the daily needs and desires of individuals across the mental health spectrum

6.6.6.4 Challenges and Considerations

- Individual differences: Recognizing the heterogeneity within the autism spectrum and tailoring CBT interventions to each individual's unique needs and strengths
- Generalization: Promoting the generalization of skills learned in therapy sessions to real-life settings and situations outside of the therapy context

In conclusion, CBT represents an encouraging and efficient intervention for the management of anxieties, social stress, and emotional regulation in children with ASD. Further research is certainly necessary to gain further insight into the long-term and effective effects of CBT; however, the present evidence offers support that CBT could well be an important treatment for individuals diagnosed with ASD.

6.6.7 Environmental Modifications

Occupational therapy often apply environmental modifications in working with people diagnosed with autism spectrum disorder (ASD) in the treatment process. The objective of such modifications is to create an environment that is more supporting yet less overpowering toward intellectual development, reducing emotional overload, and thus improving participation. Some environmental modifications Occupational therapy commonly use include:

1. *Sensory-friendly settings:* Occupational therapy may develop a healthy environment or diet that depends on the sensory need of the patient. These interventions could enhance the ability of maintaining cognitive ability in individuals with autism spectrum disorder and involve them in daily activities [72].
2. *Structured environment:* A structured environment can make a person with autism feel comfortable and minimize stress. Occupational therapy have capabilities in dealing with visual interventions, clear guidelines, and preparation strategies for people to handle the structured environment that surrounds them [73].
3. *Visual supports:* Visual supports such as charts, visual timelines, and cue cards will enable persons diagnosed with mental illness to understand expectations, changes, and outcomes. Such supports help improve communication, promote independence, and reduce stress [74].
4. *Adaptive equipment:* The occupational therpay practitioner shall recommend adaptive equipment and assistive devices to help the individual perform daily activities. These include special chairs, communication devices, and sensory aids among others in solving specific problems and improving work skills [75].
5. *Calm environment:* The development of a home, school, or clinic-based calm, comfortable, or sensory-friendly environment may allow individuals with ASD to relax and self-regulate themselves; they could then explore their environment. Occupational therapy can develop such spaces by the provision of appropriate information, lighting, and comfort seating [76].
6. *Environmental adaptation:* Occupational therapy assess and adapt environmental conditions that could interfere with the lives of a person with autism. These may include dimming light, reducing noise interference, and making activities accessible to a person in relation to the individual's capabilities and preferences.

Environmental adaptation by Occupational therapy will help individuals on the autism spectrum perform better in their environment, respond to situations in a more effective way, and be able to participate more meaningfully in everyday life.

6.6.8 Play in Occupational Therapy

Educational play is considered one of the biggest needs where children learn and think. It is supposed to be motivating and more of a great opportunity for research and development without lots of restrictions [77]. Autism is characterized by children with behavioural abnormalities, which include conditions of disorders and social and emotional problems in most cases [78]. They may struggle with motor planning and praxis, which can affect their engagement in play.

Play-based interventions are designed to address the specific needs of persons with ASD by utilizing structured play activities to facilitate learning and development. Occupational therapy use a child-centred approach, tailoring activities to the children's interests, strengths, and needs. Play activities may include sensory play, pretend play, structured games, and interactive activities [79]. Occupational therapy aim to support children with autism in developing essential skills through play, such as motor planning, social interaction, communication, and sensory integration. Sensory input plays a significant role in overall development. Occupational therapy consider sensory differences in individuals with autism and incorporate sensory integration strategies to improve play skills.

Play assessment provides valuable insights into a child's interests, abilities, and behaviours. Various standardized tests and observation tools are used to assess playfulness and inform intervention plans [80–82].

Occupational therapy employ diverse intervention approaches, including sensory integration, developmental methods like "Floortime", and even virtual reality studies [83–87]. Therapy is often child-led, with a focus on creating a supportive and enriching play environment. Parents play a vital role in supporting their development. Occupational therapy educate parents on how to facilitate play at home, considering the child's unique challenges and strengths. There is evidence that play as therapy has a positive effect on motor skills, emotional health, communication, and other developmental aspects of children with disabilities [83]. In general, play as a therapeutic activity provides a way to encourage the development and participation of children with autism disorders in important activities by allowing them to act according to their own needs.

6.6.9 Task-Oriented Treatment Approaches

Develop functional skills and promote independence through activities and participation in activities and tasks through task-oriented treatment approaches. This process is based on clinical work principles that emphasize the importance of important tasks and improve all aspects of work for people with autism spectrum disorders.

Occupational therapy for autism focuses on helping children develop the skills necessary to complete specific tasks. These approaches recognize that people with autism spectrum disorders may have problems with communication, social skills, motor skills, and cognitive skills and that these may affect their ability to function in daily life. Occupational and physical therapists help children with autism learn how to complete tasks by breaking them down into manageable steps and providing exercises.

For example, a task-oriented programme might include teaching children with autism how to brush their teeth independently. The doctor will examine your child's current abilities and any issues with this activity, such as motor coordination or friction sensation. Your child will be guided and supported at every step as they practice brushing their teeth, and their independence will gradually increase as they gain confidence and skills.

Guidance works well because it focuses on specific goals and activities that are meaningful and relevant to the child's daily life. Children with autism can learn these activities to improve their independence and quality of life. Task-oriented therapy approaches for autism can be divided into the following points:

1. *Purposeful activities:* It is important to choose activities that are meaningful and relevant to daily life. This increases engagement and motivation [22].
2. *Functional goals:* Activities are designed to achieve functional goals in a variety of areas, such as communication, social skills, motor coordination, and cognitive skills for quality work, self-care, and independent living skills [88].
3. *Structured practice:* Structured practice in a supportive environment helps develop skills. The therapist provides guidance, feedback, and adjustments as needed to encourage success [39].
4. *Task analysis:* Breaking complex tasks into manageable steps aids learning and mastery. This approach can improve the health of people with mental disorders [89].
5. *Environmental considerations:* Careful design of a clinical setting that encourages participation and interaction is important. Adaptations may include changes in attention, perception, or facial expression [90].
6. *Generalization and adaptation:* Increasing awareness of life events is an important part of therapy. Encourage people to use new skills in different situations and settings [91].
7. *Client-centred approach:* Put everyone's interests, strengths, and benefits first and promote freedom and empowerment. Treatment activities are tailored to the unique needs and goals of each person with autism [92].

By incorporating these elements into treatment, clinicians can improve the functional capacity, enhance independence, and improve the quality of life of individuals with autism.

6.6.10 Transition Planning and Vocational Rehabilitation

6.6.10.1 Transition Planning

Transition planning in autism involves preparing persons on the autism spectrum for the significant changes they will experience as they move from adolescence to adulthood. The goal is to support them in successfully transitioning to postsecondary education, employment, independent living, and community participation. Transition planning typically begins in adolescence, around the age of 14 or earlier, to allow for adequate time to identify goals, assess needs, and develop appropriate supports [93]. Transition planning is individualized and centred around the needs, strengths, preferences, and interests of the individual with autism. It involves active participation from the individual, their family, educators, service providers, and other relevant stakeholders [94].

In transition planning for individuals with ASD, Occupational therapy play a crucial role in facilitating a smooth transition from adolescence to adulthood, which are as follows:

1. *Assessment of skills and needs*: A comprehensive assessment is conducted to identify the individual's skills, abilities, challenges, interests, and aspirations in various domains, including academic, vocational, social, communication, daily living, and independent living skills [93].Occupational therapy assess the individual's strengths, challenges, and needs across various domains, including sensory processing, motor skills, social interaction, communication, activities of daily living, vocational skills, and independent living skills[95].
2. *Collaboration in transition team*: Occupational therapy collaborate with other members of the transition team, including educators, vocational counsellors, speech therapists, psychologists, and parents, to develop a comprehensive transition plan [94].
3. *Individualized transition goals*: Based on assessment verdicts, Occupational therapy help develop individualized transition goals that focus on enhancing functional abilities and preparing for postsecondary education, employment, and independent living [95].
4. *Skill development and training*: Occupational therapy provide skill development and training in areas such as self-care, organization, time management, problem-solving, decision-making, social skills, vocational skills, and adaptive behaviours [94].
5. *Environmental modifications*: Occupational therapy identify environmental barriers and recommend modifications to support the individual's participation and success in various settings, including school, work, home, and community environments [96].
6. *Transition preparation programmes*: Occupational therapy may lead or participate in transition preparation programmes that provide opportunities for hands-on learning, vocational exploration, job shadowing, internships, and community-based experiences [94].
7. *Assistive technology and accommodations*: Occupational therapy assess the individual's need for assistive technology and accommodations in academic and vocational settings, providing recommendations and training as needed [96].
8. *Family education and support*: Occupational therapy offer education and support to families on topics related to transition planning, including understanding the individual's rights, accessing community resources, and promoting independence and self-advocacy skills [95]

6.6.10.2 Vocational Rehabilitation

Vocational rehabilitation plays a vital role in supporting persons with ASD in finding and maintaining employment. The key points of this are the following:

1. *Employment challenges*: Many persons with autism face difficulties in securing and retaining jobs, with a significant portion remaining unemployed [97]. Challenges often stem from deficits in social and communication skills, which are essential for workplace success [98].
2. *Types of employment opportunities*: Vocational opportunities for individuals with autism can fall into two main categories: integrated, competitive employment with supportive opportunities and independent, non-competitive settings such as work activity centres. Supportive employment involves long-term support provided through a work coaching system [99].
3. *Supportive employment programmes*: Research has shown that supportive employment programmes are effective in facilitating workforce inclusion of individuals with autism. These programmes involve multidisciplinary teams that assess, train, and support individuals in the workplace [100].
4. *Role of occupational therapy*: Occupational therapy play various roles throughout the vocational rehabilitation process. They evaluate sensory and motor skills, provide training and adaptations for work tasks, and address cognitive and social skills necessary for employment success [101].
5. *Assessment*: Vocational assessment begins with interviews, self-reports, and behavioural observations to identify individual strengths, interests, and needs [102]. Occupational therapy assess sensory-motor skills, socialization abilities, cognitive skills, and participation in daily activities using standardized assessments [103–105].
6. *Intervention*: Occupational therapy train individuals in abilities required for their occupation and educate natural supports, for instance, co-workers and family members, on adaptive techniques and environmental adaptations. They also address complex cognitive skills and sensory needs through various strategies and technologies [106].
7. *Assistive technology*: Assistive technology, including computer-based systems, video modelling, and personal digital assistants, can enhance natural support for persons with autism in the workplace. Occupational therapy can utilize technology to support sensory needs and improve productivity on the job [107].

In summary, vocational rehabilitation for individuals with autism involves comprehensive assessment, skill training, and ongoing support to facilitate successful employment outcomes. Occupational therapy play a vital role in addressing the unique needs of persons with autism and promoting their integration into the workforce.

6.6.11 Visual Supports and Strategies

Visual supports and strategies play an important role in improving communication, behaviour management, social skills, emotional regulation, and overall quality of life for people with mental illness. Occupational therapy use visual supports and quantitative strategies to improve communication, behaviour management, daily living, and social interactions among patients with ASD. Some visual supports and strategies for ASD include:

1. *Visual timeline:* A visual timeline that uses pictures, symbols, or words to describe the day's work. They provide models and predictions that help people with autism understand and predict change [108].
2. *Visual timers:* Visual timers use visual cues, such as changing colour or moving symbols, to indicate the passage of time. They help people with autism manage time, increase awareness of time constraints, and encourage flexibility in play [109].
3. *Visual task analysis:* Visual task analysis breaks down complex tasks or activities into smaller, sequential steps using pictures, symbols, or written instructions. It helps persons with ASD understand and complete tasks independently by providing visual guidance [110].
4. *Visual supports for social skills:* Visual supports, such as social stories, video modelling, and comic strip conversations, help persons with ASD understand social situations, interpret social cues, and learn appropriate social behaviours [111].
5. *Visual supports for communication:* Visual supports, including picture exchange systems (PECS), communication boards, and visual choice boards, aid persons with ASDin expressing their needs, preferences, and desires [112].
6. *Visual supports for behaviour management:* Visual supports, such as behaviour charts, token boards, and visual rules, assist persons with autism spectrum disorder in understanding expectations, monitoring their behaviour, and earning rewards for positive behaviour [109].
7. *Visual supports for sensory regulation:* Visual supports, such as sensory break cards, sensory cue cards, and visual calm-down kits, help persons with ASD identify sensory triggers, self-regulate their emotions, and access sensory accommodations [112].
8. *Environmental visual supports:* Environmental visual supports, such as labels, signs, and visual boundaries, help persons with ASD navigate and organize their environment, understand spatial concepts, and reduce sensory overload [110].
9. *Individualized visual supports:* Visual supports should be individualized to meet the specific needs, preferences, and abilities of each person with ASD. Occupational therapy collaborate with individuals and their families to select and create visual supports that are meaningful and effective [108].
10. *Education and training:* Occupational therapy provide education and training to individuals with autism, their families, caregivers, teachers, and other

professionals on the use and integration of support in daily living and activities [111].

In conclusion, cognitive supports and strategies are important tools that therapists use to promote independence, communication, socialization, behaviour management, and emotional regulation in individuals with autism.

6.6.12 Family Education and Support

Occupational therapy (OT) plays an important role in providing family education and support to individuals with autism. Some of the important roles of OT in this regard are:

1. *Autism spectrum disorder education:* Practitioners provide families with information about autism, including its characteristics, developmental impacts, and evidence-based interventions. Helping families understand the unique needs and challenges of individuals with autism allows them to make informed decisions about the care and support of their loved ones [113].
2. *Understanding sensory perception:* Many people with ASD have hearing issues that can affect their daily activities. Therapists work to educate families about sensory issues and how they can present in people on the autism spectrum. They provide ideas and strategies to create a sense of well-being in the environment and promote emotional control in the home [114].
3. *Teaching activities of daily living:* Therapists work with families to teach activities of daily living such as personal care, hygiene, meal preparation, and administrative control at home. They provide strategic and adaptive strategies to promote independence and participation in daily activities [115].
4. *Developing communication and social skills:* Communication and social skills are problems experienced by people with mental disorders. Occupational therapy work with families to implement strategies to improve communication and social relationships. They may recommend supplementary and alternative communication (AAC), community stories, visual aids, and social skill training [113].
5. *Behaviour management strategies:* Occupational therapy assist families in developing effective behaviour management strategies to address challenging behaviours commonly associated with ASD. They guide positive reinforcement techniques, visual schedules, behaviour modification plans, and sensory-based interventions to promote self-regulation and reduce stress [116].

By offering education and support, occupational therapy empower families to well understand and meet the needs of their loved ones with ASD, ultimately enhancing the overall welfare and life quality for the individual and their family.

6.6.13 Collaboration with Other Professionals

Collaboration among professionals is fundamental and essential in the effective management of autism spectrum disorders (ASDs), and occupational therapy (OTs) play a pivotal role in interdisciplinary teams. OTs most often collaborate with other professionals like:

1. *Speech-language pathologists (SLPs)*: OTs and SLPs often collaborate closely to address communication and sensory challenges in persons with autism spectrum disorder. They collaborate on strategies to improve social communication skills, feeding difficulties, and sensory processing challenges and also work together to develop comprehensive intervention plans that target both speech and language skills and sensory processing difficulties [117].
2. *Behaviour therapist/analysts:* Behaviour analysts focus on modifying behaviour, while OTs focus on integrating sensory and driving skills. Collaboration between behaviour analysts and OTs ensures that behaviour plans meet needs and encourage participation in meaningful activities [118].
3. *Psychologist/psychiatrist:* Psychologists and psychiatrists provide assessment, diagnosis, and mental health treatment for individuals with autism. OTs work with students to develop holistic intervention plans that meet psychological and functional needs.
4. *Educators/teachers*: Collaboration between OTs and teachers is important to create a meaningful classroom environment and to create individualized programmes (IEPs) for students with ASD. OTs provide strategies to promote academic and social participation [117].
5. *Physical therapist (PT)*: OTs and PTs work together to address motor coordination, balance, and gross motor skill issues in individuals with disabilities. Together, they create a comprehensive intervention plan for fines and gross motor deficits [118].
6. *Therapists/medical professionals:* Occupational therapy work with paediatricians, psychiatrists, and other healthcare professionals to provide comprehensive care for individuals with autism. They provide information on functional capacity, sensory issues, and mediating factors.
7. *Parents/caregivers:* The involvement of parents and caregivers is important in the treatment of autism spectrum disorder. OTs provide education, training, and support to help families understand their child's sensory and motor needs and implement strategies to promote independence and participation in daily activities [117].
8. *Social workers*: Social workers play a crucial role in connecting families with community resources and support services. OTs collaborate with social workers to address the social and emotional needs of persons with autism spectrum disorder and their families [118].

Early detection and interventions are crucial for improving outcomes for persons with autism spectrum disorder. With better understanding and support, people with autism may lead satisfying lives and contribute meaningfully to society.

Here some key features of ASD, such as abnormal social interaction, self-focus, and impairments in connectivity and synaptogenesis, have been highlighted before the discussion about the occupational therpay practitioner role and strategies as understanding these characteristics is essential for developing effective interventions and support systems for persons with autism spectrum disorder and their families.

6.7 Conclusion

Ultimately, occupational therapy is an important foundation in the recovery experience of someone having ASD. Occupational therapy can develop a broad variety of personalized interventions that respond to the particular needs of each individual by recognizing and responding to essential features of ASD which includes self-focus, deviant social interactions, and poor connectivity and synaptogenesis.

As part of multidisciplinary treatment teams, Occupational therapy ensure that all aspects of the treatment plan are integrated effectively in order to foster development and functional skills in children with ASD. Clinicians should be aware of evidence-based practices in today's clinical world while endeavouring to craft effective intervention strategies using their clinical skills and knowledge of the particular child.

Enhancing quality of life for individuals with autism spectrum disorder is a primary objective in occupational therapy, enabling them to engage actively in daily activities. When integrated into holistic treatment plans that prioritize the client, sensory integration therapy significantly benefits those facing sensory processing difficulties and supports their overall well-being.

The treatment strategies and techniques used in OT interventions assist individuals with ASD in developing essential skills needed to effectively carry out various life roles and tasks. Occupational therapy work to empower individuals with ASD to reach their full potential through a comprehensive approach that emphasizes evidence-based practices, collaborative teamwork, and client-centred care.

References

1. DeLany J. Scope of occupational therapy services for individuals with autism spectrum disorder across the life course. Am J Occup Ther. 2015;69(Suppl 3):6913410054p1-6913410054p12.
2. American Psychiatric Association. Neurodevelopmental disorders: diagnostic and statistical manual of mental disorders DSM-5TM. 5th ed. Arlington: American Psychiatric Association; 2013. p. 50–9.
3. Zeidan J, Fombonne E, Scorah J, Ibrahim A, Durkin MS, Saxena S, Yusuf A, Shih A, Elsabbagh M. Global prevalence of autism: a systematic review update. Autism Res. 2022;15(5):778–90.

4. Rudra A, Belmonte MK, Soni PK, Banerjee S, Mukerji S, Chakrabarti B. Prevalence of autism spectrum disorder and autistic symptoms in a school-based cohort of children in Kolkata, India. Autism Res. 2017;10(10):1597–605.
5. Raina SK, Chander V, Bhardwaj AK, Kumar D, Sharma S, Kashyap V, Singh M, Bhardwaj A. Prevalence of autism spectrum disorder among rural, urban, and tribal children (1–10 years of age). J Neurosci Rural Pract. 2017;8(3):368–74.
6. Newschaffer CJ, Croen LA, Daniels J, Giarelli E, Grether JK, Levy SE, Mandell DS, Miller LA, Pinto-Martin J, Reaven J, et al. The epidemiology of autism spectrum disorders. Annu Rev Public Health. 2007;28:235–58.
7. Center for Disease Control and Prevention. Autism. Morbidity and mortality weekly report surveillance summaries March 28, 2014;63(2).
8. Brentani H, Paula CS, Bordini D, et al. Autism-spectrum-disorders: an overview on diagnosis and treatment. Rev BrasPsiquiatria. 2013;35(Suppl 1):S62–72.
9. Lord C, Risi S, Lambrecht L, Cook EH Jr, Leventhal BL, DiLavore PC, Pickles A, Rutter M. The autism diagnostic observation schedule-generic: a standard measure of social and communication deficits associated with the spectrum of autism. J Autism Dev Disord. 2000;30(3):205–23.
10. Patra S, Arun P. Use of Indian scale for assessment of autism in child guidance clinic: an experience. Indian J Psychol Med. 2011;33(2):217–9.
11. Juneja M, Mishra D, Russell PS, Gulati S, Deshmukh V, Tudu P, et al. INCLEN diagnostic tool for ASD (INDT-ASD): development and validation. Indian Pediatr. 2014;51:359–65.
12. Johnson CP, Myers SM. American Academy of Pediatrics Council on Children with Disabilities. Identification and evaluation of children with autism spectrum disorders. Pediatrics. 2007;120(5):1183–215.
13. American Occupational Therapy Association. Occupational therapy practice framework: domain and process, 3rd ed. Am J Occup Ther. 2014;68(Suppl 1):S1–S48.
14. Pekçetin S, Günal A. Early intervention in pediatric occupational therapy. In: Huri M, editor. Occupational therapy. IntechOpen; 2017. https://doi.org/10.5772/intechopen.68316.
15. Bumin G, Huri M, Salar S, Kayihan H. Occupational therapy in autism. In: Fitzgerald M, editor. Autism spectrum disorder. IntechOpen; 2015. https://doi.org/10.5772/59168.
16. Lane AE, Young RL, Baker AE, Angley MT. Sensory processing subtypes in autism: association with adaptive behavior. J Autism Dev Disord. 2010;40(1):112–22.
17. Case-Smith J, Weaver LL, Fristad MA. A systematic review of sensory processing interventions for children with autism spectrum disorders. Autism. 2015;19(2):133–48.
18. Ayres AJ. Sensory integration and learning disorders. Los Angeles: Western Psychological Services; 1972.
19. Watling R, Deitz J, Kanny EM, McLaughlin JF. Current practice of occupational therapy for children with autism. Am J Occup Ther. 1999;53(5):498–505.
20. Schaaf RC, Miller LJ. Occupational therapy using a sensory integrative approach for children with developmental disabilities. Ment Retard Dev Disabil Res Rev. 2005;11(2):143–8.
21. Kimball JG. Sensory integration frame of reference: theoretical base function/dysfunction continua, and guide to evaluation. In: Kramer P, Hinojosa J, editors. Frames of reference of pediatric occupational therapy; 1999. p. 119–59.
22. Case-Smith J, Arbesman M. Evidence-based review of interventions for autism used in or of relevance to occupational therapy. Am J Occup Ther. 2008;62(4):416–29.
23. Baranek GT, et al. Sensory features and repetitive behaviors in children with autism and developmental delays. Autism Res. 2015;8(5):471–80.
24. Pfeiffer BA, et al. Effectiveness of sensory integration interventions in children with autism spectrum disorders: a pilot study. Am J Occup Ther. 2011;65(1):76–85.
25. Ingersoll B, Schreibman L, Stahmer A. Brief report: differential treatment outcomes for children with autistic spectrum disorder based on level of peer social avoidance. J Autism Dev Disord. 2001;31(3):343–9.
26. Baron-Cohen S, Leslie AM, Frith U. Does the autistic child have a "theory of mind?". Cognition. 1985;21(1):37–46.

27. Tager-Flusberg H, Paul R, Lord C. Language and communication in autism. In: Handbook of autism and pervasive developmental disorders, vol 1. 2005. p. 335–64.
28. Ozonoff S. Executive functions in autism. In: Learning and cognition in autism. Springer; 1995. p. 199–219.
29. Pennington BF, Ozonoff S. Executive functions and developmental psychopathology. J Child Psychol Psychiatry. 1996;37(1):51–87.
30. Happé F, Briskman J, Frith U. Exploring the cognitive phenotype of autism: Weak "central coherence" in parents and siblings of children with autism: I. Experimental tests. J Child Psychol Psychiatry. 2001;42(3):299–307.
31. Kasari C, Locke J, Gulsrud A, Rotheram-Fuller E. Social networks and friendships at school: Comparing children with and without ASD. J Autism Dev Disord. 2011;41(5):533–44.
32. Leaf JB, Taubman M, Bloomfield S, Palos-Rafuse L, Leaf R, Mc Eachin J, Oppenhaim ML. Increasing social skills and pro-social behavior for three children diagnosed with autism through the use of a reaching package. Res Autism Spectr Disord. 2009;3(1):275–89.
33. Matson JL, Matson ML, Rivet TT. Social-skills treatments for children with autism spectrum disorders. An overview. Behav Modif. 2007;31(5):682–707.
34. Attwood T. Strategies for improving the social integration of children with Asperger syndrome. Autism Int J Res Practice. 2000;4(1):85–100.
35. Baker JE. Social skills training for children and adolescents with Asperger Syndrome and Social Communication Problems. Shawnee Mission: Autism Asperger Publishing Company; 2003.
36. Myles B. Overview of Asperger syndrome. In: Baker JE, editor. Social skills training. Autism Asperger Publishing Company; 2003. p. 9–17.
37. Howlin P, Baron-Cohen S, Hadwin J. Teaching children with autism to mind-read: a practical guide. Chichester: Willey; 2003.
38. Rao PA, Beidel DC, Murray MJ. Social skills interventions for children with Asperger's syndrome or high-functioning autism: a review and recommendations. J Autism Dev Disord. 2008;38(2):353–61.
39. Rogers SJ, Vismara LA. Evidence-based comprehensive treatments for early autism. J Clin Child Adolesc Psychol. 2008;37(1):8–38.
40. Rodgers J. Visual perception and Asperger syndrome: central coherence deficit or hierarchization deficit? Int J Res Pract. 2000;4(3):321–9.
41. Reynhout G, Carter M. Social Stories™ for children with disabilities. J Autism Dev Disord. 2006;36(4):445–69.
42. DiSalvo CA, Oswald DP. Peer-mediated interventions to increase the social interaction of children with autism consideration of peer expectancies. Focus Autism Other Dev Disabil. 2002;17(4):198–207.
43. Cartledge G. Teaching social skills to children: innovative approaches, vol 89. Allyn & Bacon; 1986.
44. Roley SS, DeLany JV, Barrows CJ, Brownrigg S, Honaker D, Sava DI, et al. American occupational therapy association commission on practice. Occupational therapy practice framework: domain & practice, 2nd edition. Am J Occup Ther. 2008;62(6):625–83.
45. LeVesser P, Berg C. Participation pattern in preschool children with autism spectrum disorder. OTJR Occup Particip Health. 2011;31(1):33–9.
46. Rodger S, Umaibalan V. The routines and rituals of families of typically developing children compared with families of children with autism spectrum disorder: an exploratory study. Br J Occup Ther. 2011;74(1):20–6.
47. Gillham JE, Carter AS, Volkmar FR, et al. Toward a developmental operational definition of autism. J Autism Dev Disord. 2000;30(4):269–78.
48. Kenworthy L, Black DO, Harrison B, et al. Are executive control functions related to autism symptoms in high-functioning children? Child Neuropsychol. 2009;15(5):425–40.
49. Liss M, Harel B, Fein D, et al. Predictors and correlates of adaptive functioning in children with developmental disorders. J Autism Dev Disord. 2001;31(2):219–30.

50. Shepherd J. Activities of daily living and adaptations for independent living. In: Case Smith J, editor. Occupational therapy for children. Elsevier Mosby; 2005. p. 356–409.
51. Sparrow SS, Balla DA, Cicchetti DV. Vinel and adaptive behavior scales: interview edition, survey form manual. Circle Pines: American Guidance Service; 1984.
52. Haley SM. Pediatric Evaluation of Disability Inventory (PEDI): development, standardization, and administration manual. PEDI Research Group; 1992.
53. Schopler E, Mesibov GB, Hearsey K. Structured teaching in the TEACCH system. In: InSchopler E, Mesibov GB, editors. Learning and cognition in autism. New York: Plenum; 1995. p. 243–67.
54. Simpson RL, Myles BS, Ganz JB. Efficacious interventions and treatments for learners with autism spectrum disorders. In: Simpson RL, Myles BS, editors. Educating children and youth with autism: strategies for effective practice. 2nd ed. Austin: Pro-Ed, Inc.; 2008. p. 477–512.
55. Bryan LC, Gast DL. Teaching on-task and on-schedule behaviors to high-functioning children with autism via Picture activity schedules. J Autism Dev Disord. 2000;30:553–67.
56. Van Laarhoven T, Kraus E, Karpman K, Nizzi R, Valentino J. A comparison of picture and video prompts to teach daily living skills to individuals with autism. Focus Autism Other Dev Disabil. 2010;25(4):195–208.
57. Goldsmith TR, LeBlanc LA. Use of technology in interventions for children with autism. J Early Intensive Behav Intervent. 2004;1(2):166–78.
58. Shipley-Benamou R, Lutzker JR, Taubman M. Teaching daily living skills to children with autism through instructional video modelling. J Posit Behav Interv. 2002;4(3):166–77.
59. Kuo R, Chand M, Lyu CW, Heh JS. A pilot study of the situated game for autistic children learning activities of daily living. Res Pract Technol Enhanc Learn. 2013;8(2).
60. Schaaf RC, Roley SS. Sensational ways to play. In: Schaaf RC, Roley SS, editors. Sensory integration: applying clinical reasoning to practice with diverse populations. Pro. ed; 2006. p. 195–242.
61. Chiesa M. Radical behaviourism: the philosophy and the science. Boston: Authors Cooperative; 2004.
62. Ryan JB, Hughes EM, Katsiyannis A, McDaniel M, Sprinkle C. Research-based educational practices for students with autism spectrum disorders. Teach Except Child. 2011;43(3):56–64.
63. Bruce MA, Borg B. Psychosocial frames of reference: core for occupation-based practice. Slack Incorporated; 2002.
64. Association for Behavioral and Cognitive Therapies Evidenced-based mental health treatments for children and adolescence; 2010.
65. Ikiugu MN. Instrumentalism in occupational therapy: an argument fora pragmatic conceptual model of practice. Int J Psychosoc Rehabil. 2004;8:109–17.
66. Kielhofner G. Conceptual foundations of occupational therapy practice. FA Davis; 2009.
67. Haglund L, Ekbladh E, Thorell LH, Hallberg IR. Practice models in Swedish psychiatric occupational therapy. Scand J Occup Ther. 2000;7(3):107–13.
68. Buron KD, Curtis M. The incredible 5-point scale. Shawnee Mission; 2003.
69. Gagnon E. Power cards: using special interests to motivate children and youth with asperger syndrome and autism. Shawnee Mission: Autism Asperger Publishing Co.; 2001.
70. Bellini S, Akullian J. A meta-analysis of video modelling and video self-modelling interventions for children and adolescents with autism spectrum disorders. Except Child. 2007;73(3):264–87.
71. Mailloux Z, May Benson TA, Summers CA, Miller LJ, Brett Green B, Burke JP. Goal Attainment scaling as a measure of meaningful outcomes for children with sensory integration disorders. Am J Occup Ther. 2007;61(2):254–9.
72. Schaaf RC, Mailloux Z. Sensory integration and praxis patterns in children with autism. Am J Occup Ther. 2015;69(1):6901220020.
73. Myles BS, Cook KT, Miller NE, Rinner L, Robbins LA. Asperger syndrome, and sensory issues: practical solutions for making sense of the world. 2004;39(1):283–90.
74. Grandin T. How does visual thinking work in the mind of a person with autism? A personal account. Philos Trans R Soc Lond Ser B Biol Sci. 2009;364(1522):1437–42.

75. Watling RL, Dietz J. Immediate effect of Ayres's sensory integration-based occupational therapy intervention on children with autism spectrum disorders. Am J Occup Ther. 2007;61(5):574–83.
76. Myles BS, Southwick J. Asperger syndrome and difficult moments: practical solutions for tantrums, rage, and meltdowns. New York: Autism Asperger Publishing Co; 2005.
77. Bundy AC. Play theory and sensory integration. In: Sensory integration: theory and practice. 1991. p. 46–68.
78. Pierce-Jordan S, Lifter K. Interaction of social and play behaviors in preschoolers with and without pervasive developmental disorder. Top Early Child Spec Educ. 2005;25(1):34–47.
79. Boutot EA, Guenther T, Crozier S. Let's play: teaching play skills to young children with autism. Educ Train Dev Disabil. 2005;40(3):285–92.
80. Pellegrini AD, Dupuis D, Smith PK. Play in evolution and development. Dev Rev. 2007;27(2):261–76.
81. Bundy AC, Nelson L, Metzger M, Bingaman K. Validity and reliability of a test of playfulness. Occup Ther J Res. 2001;21(4):276–92.
82. Knox S. Development and current use of the Knox Preschool Play Scale. In: Play in occupational therapy for children. 1997. p. 35–51.
83. Yack E, Sutton S, Aquilla P. Building bridges through sensory integration: therapy for children with autism and other pervasive developmental disorders. Future Horizons; 2003.
84. Anzalone ME, Williamson GG. Sensory processing and motor performance in autism spectrum disorders. Autism Spectrum Disord Trans Dev Perspect. 2000;9:143–66.
85. Pajareya K, Nopmaneejumruslers K. A pilot randomized controlled trial of DIR/Floortime™ parent training intervention for pre-school children with autistic spectrum disorders. Autism. 2011;15(5):563–77.
86. Solomon R, Necheles J, Ferch C, Bruckman D. Pilot study of a parent training program for young children with autism: the PLAY Project home consultation program. Autism. 2007;11(3):205–24.
87. Strickland D. Virtual reality for the treatment of autism. In: Studies in health technology and informatics. 1997. p. 81–6.
88. Schaber P. Evaluation and treatment of autistic disorders: applied behavior analysis, evidence-based practice, and the controversy over their use. Am J Occup Ther. 2001;55(1):78–82.
89. Ploog BO. Teaching a student with ASD to write persuasive essays using a structured writing intervention. Focus Autism Other Dev Disabil. 2010;25(3):145–57.
90. Miller-Kuhaneck H, Watling R, Kuhaneck M, Henry D. Sensory integration: an applied view of sensory dysfunction. In: Pediatric occupational therapy handbook: a guide to diagnoses and evidence-based interventions. 2001. p. 261–92.
91. Odom SL, Boyd BA, Hall LJ, Hume K. Evaluation of comprehensive treatment models for individuals with autism spectrum disorders. J Autism Dev Disord. 2010;40(4):425–36.
92. Law M, Baptiste S, Carswell A, McColl MA, Polatajko H, Pollock N. Canadian occupational performance measure. Ottawa: CAOT Publications ACE; 2014.
93. Shogren KA, Plotner AJ, Curtis MJ. Transition assessment and planning for youth with severe intellectual and developmental disabilities. Baltimore: Paul H Brookes Publishing; 2015.
94. Van Laarhoven T, Kraus L, Powell S. Occupational therapy's role in transition planning for students with autism spectrum disorder. OT Pract. 2016;21(17):10–4.
95. Hillier A, Fish T, Cloppert P, Beversdorf DQ. Outcomes of a social and vocational skills support group for adolescents and young adults on the autism spectrum. Focus Autism Other Dev Disabil. 2007;22(2):107–15.
96. Barnhill GP, Hagiwara T, Myles BS, Simpson RL, Brick ML. The effectiveness of social skills intervention targeting nonverbal communication for adolescents with Asperger syndrome and related pervasive developmental delays. Focus Autism Other Dev Disabil. 2002;17(2):112–8.
97. Wehman P, Lau S, Molinelli A, Brooke V, Thompson K, Moore C, West M. Supported employment for young adults with autism spectrum disorder: preliminary data. Res Pract Persons Severe Disabil. 2012;37:160–9.

98. Allen KD, Burke RV, Howard MR, Wallace DP, Bowen SL. Use of audio cuing to expand employment opportunities for adolescents with autism spectrum disorders and intellectual disabilities. J Autism Dev Disord. 2012;42(11):2410–9.
99. Orentlicher ML, Olson LJ. Transition from school to adult life for students with an autism-spectrum-disorder. In: Kuhaneck MH, editor. Autism, a comprehensive occupational therapy approach. New York: The American Occupational Therapy Association; 2004. p. 359–86.
100. Reed CA, Rumrill PD Jr. Supported employment: principles and practices for interdisciplinary collaboration. Work. 1997;9(3):237–44.
101. Pitts DB. Work as occupation. In: Brown C, Stoffel VC, editors. Occupational therapy in mental health: a vision for participation. Philadelphia: F.A. Davis Company; 2011.
102. Taylor JL, Seltzer MM. Developing a vocational index for adults with autism spectrum disorders. J Autism Dev Disord. 2012;4:2669–79.
103. Brown C, Dunn W. Adolescent/adult sensory profile. San Antonio: Psychological Corporation; 2002.
104. Gal E, Meir AB, Katz N. Development and reliability of the autism work skills questionnaire (AWSQ). Am J Occup Ther. 2013;67(1):e1–5.
105. Capo LC. Autism, employment, and the role of occupational therapy. Work. 2001;16(3):201–7.
106. Wilczynski S, Trammell B, Clarke L. Improving employment outcomes among adolescents and adults on the autism spectrum. Psychol Sch. 2013;50(9):876–87.
107. Smith TL, Segal R, Segal J. Using visual supports with infants and toddlers with autism spectrum disorder. Young Exceptional Child. 2015;18(3):16–25.
108. Crosland KA, Dunlap G. Providing visual support to students with autism spectrum disorder. Teach Except Child. 2015;47(4):207–15.
109. Kellems RO, Morningstar ME. Using visual supports to facilitate transitions for students with autism spectrum disorder. Teach Except Child. 2016;48(1):23–31.
110. Ramdoss S, Lang R, Fragale C, Britt C, O'Reilly M, Sigafoos J, et al. Use of computer-based interventions to teach communication skills to children with autism spectrum disorders: a systematic review. J Behav Educ. 2012;21(1):55–77.
111. Preston D, Carter M, Lewin J. A review of the research on strategies for managing impulsive aggression when caring for persons with developmental disabilities. J Policy Pract Intellect Disabil. 2014;11(2):81–9.
112. American Occupational Therapy Association. Autism spectrum disorder. 2020. Retrieved from: https://www.aota.org/Practice/Children-Youth/Autism.aspx
113. Roley SS, Mailloux Z, Parham LD, Schaaf RC, Lane CJ, Cermak S. Sensory integration and processing in children with autism: a review of the literature. In: Comprehensive guide to autism. Cham: Springer; 2015. p. 245–60.
114. Watling R, Deitz J, Kanny EM. Occupational therapy practice guidelines for children and adolescents with challenges in sensory processing and sensory integration. AOTA Press; 2018.
115. Howlin P. Social communication intervention for children with autism: an overview. In: Handbook of autism and pervasive developmental disorders. Wiley; 2013. p. 881–905.
116. Case-Smith J, O'Brien JC. Occupational therapy for children and adolescents. 7th ed. Elsevier Health Sciences; 2015.
117. Watling R, Hauer S. Effectiveness of Ayres Sensory Integration® and sensory-based interventions for people with autism spectrum disorder: a systematic review. Am J Occup Ther. 2015;69(5):6905180030p1–6905180030p10.
118. American Occupational Therapy Association. Occupational therapy practice framework: domain and process (3rd ed.). Am J Occup Ther. 2018;72(Suppl. 2):7212410010.

Aquatic Therapy and Autism: A Therapeutic Alliance

7

Riya Kalra, Subhashish Chatterjee, Manu Goyal, and Kanu Goyal

7.1 Introduction to Aquatic Therapy and Autism

Aquatic therapy utilizes water-based exercises and methods to promote both physical and mental well-being. It harnesses water's physical properties like buoyancy, resistance, and hydrostatic pressure to establish a therapeutic setting that is particularly advantageous for individuals with diverse conditions, including autism spectrum disorder (ASD) [1, 2].

Autism spectrum disorder (ASD) is a developmental condition identified by difficulties in social skills, repetitive behaviors, and communication. People with ASD may also encounter sensory processing challenges, exhibiting either heightened sensitivity or reduced responsiveness to sensory stimuli [3, 4].

7.1.1 Overview of Autism Spectrum Disorder (ASD)

Autism spectrum disorder is a developmental disability marked by significant social, communication, and behavioral challenges. The manifestation of ASD varies widely among individuals, hence the term "spectrum." Symptoms typically appear in early childhood and affect daily functioning. The Centers for Disease Control and Prevention (CDC) highlights that ASD is characterized by repetitive behaviors and difficulty in social interaction. Diagnosis involves assessing the child's behavior and development, with early intervention being crucial for improving outcomes [5].

R. Kalra · S. Chatterjee · M. Goyal · K. Goyal (✉)
Maharishi Markandeshwar Institute of Physiotherapy and Rehabilitation, Maharishi Markandeshwar (Deemed to be University), Mullana - Ambala, Haryana, India
e-mail: subhasishphysio@mmumullana.org; manu.goyal@mmumullana.org; kanu.goyal@mmumullana.org

M. Goyal, K. Goyal (eds.), *Rehabilitation Approach in Autism*,
https://doi.org/10.1007/978-981-96-4162-8_7

7.1.2 Importance of Alternative Therapies in ASD Management

The management of ASD is multifaceted, incorporating behavioral interventions, educational strategies, medication, and supportive therapies tailored to the individual's needs. Alternative therapies, such as aquatic therapy, music therapy, dance therapy, acupuncture, and yoga, have gained attention for their potential to complement traditional treatments [6, 7]. These therapies often focus on improving quality of life, reducing symptoms, and supporting overall development. For individuals with ASD, aquatic therapy can be particularly beneficial due to its sensory and physical advantages. The sensory experience of being in water can help in regulating sensory overload, a common issue in ASD. Moreover, the social and playful aspects of aquatic therapy can enhance communication and interaction skills in a less structured environment than typical therapy sessions [7].

7.1.3 Need for Personalized Therapeutic Approaches in Autism Care

Considering the wide range of characteristics within the autism spectrum, individualized therapeutic approaches are crucial for providing effective care. Personalized therapy involves customizing interventions to address the specific needs, strengths, preferences, and challenges of each person with ASD. This approach may involve a variety of therapeutic methods, such as behavioral therapies, speech and language therapy, occupational therapy, and complementary treatments like aquatic therapy. The aim is to promote the individual's development, improve their quality of life, and help them navigate social and environmental situations more effectively [8, 9].

7.2 Definition of Aquatic Therapy

Aquatic therapy is a type of physical therapy conducted in a water-based environment, typically in a comfortably heated pool. The therapeutic benefits of water, including buoyancy, resistance, and hydrostatic pressure, contribute to its effectiveness in treating various conditions. Buoyancy reduces the impact of gravity, enabling greater joint mobility and decreased pain during movements. Resistance aids in muscle strengthening, while hydrostatic pressure can alleviate swelling and enhance circulation. These properties render aquatic therapy beneficial for individuals with diverse conditions, including neurological disorders, musculoskeletal issues, and developmental disabilities [10].

7.2.1 Theoretical Foundations of Aquatic Therapy for Autism

Aquatic therapy's therapeutic benefits are based on sensory integration theory, which suggests that difficulties in processing sensory information can significantly impact behavior and skill development. The unique aquatic environment can aid individuals with ASD in processing sensory information in a more manageable way. Water's buoyancy reduces gravity's effects, allowing individuals with physical limitations to move more easily. Water resistance can enhance muscle tone and strength, while hydrostatic pressure can improve proprioceptive feedback, helping individuals better perceive their body's position in space. Additionally, water's thermal conductivity can relax muscles and improve circulation [5, 11, 12].

7.2.2 Benefits of Aquatic Therapy for Children with Autism

Aquatic therapy offers a range of benefits for children with autism, affecting various domains including physical, sensory, cognitive, emotional, and social aspects. Physically, it can improve motor skills and coordination. Sensory and cognitive benefits involve enhanced sensory processing and concentration. Emotionally, the soothing nature of water can alleviate stress and boost self-esteem. Socially, group sessions can promote interactions and communication skills. Researches support these benefits, showing positive outcomes in social interaction, daily activity performance, and overall quality of life for children with autism engaging in aquatic therapy programs [7, 13, 14].

7.2.3 Implementing Aquatic Therapy for Autistic Children

Implementing aquatic therapy for children with autism involves assessing the child's unique needs and preferences to design a tailored program. Goals may include improving motor skills, enhancing sensory processing, or fostering social interactions. Activities should be chosen based on the child's interests to increase motivation and engagement, with safety as a paramount concern. Aquatic therapy sessions should be planned with flexibility, allowing adaptations based on the child's response. Engagement strategies might include the use of visual aids, structured routines, and positive reinforcement to ensure a positive and productive experience [15, 16].

7.2.4 Enhanced Sensory Integration Through Aquatic Therapy

Children with autism often experience challenges with sensory integration, where the brain struggles to effectively process and respond to sensory stimuli. Aquatic therapy can play a pivotal role in addressing these sensory processing issues [12]. The unique sensory environment of water—its pressure, temperature, and resistance—offers a controlled setting for sensory stimulation. In water, children can experience a range of sensory inputs in a manner that is often perceived as more predictable and less overwhelming than similar stimuli on land. Buoyancy provides a sense of weightlessness and can decrease sensory overload, making it easier for a child to focus on processing new sensory information. The hydrostatic pressure exerted by water is evenly distributed across the body, providing proprioceptive feedback, which can help a child with autism understand their body's position in space, a common challenge in ASD [17, 18].

7.2.5 Social and Emotional Benefits

Aquatic therapy offers significant social and emotional benefits for children with autism. The water environment promotes interactions that can be challenging in other settings. Group activities in the pool support the development of social skills like turn-taking, following instructions, and cooperative play in a noncompetitive atmosphere focused on participation and enjoyment rather than winning, which helps reduce anxiety and stress.

Emotionally, the calming effect of water coupled with the achievement of new skills can enhance self-esteem and confidence. Children with autism, who may face difficulties with conventional sports and social activities, often find aquatic therapy to be a rewarding and positive experience [16, 17].

7.3 Implementing Aquatic Therapy: Tailored Approaches

When implementing aquatic therapy for children with autism, it is essential to use a personalized approach that takes into account each child's unique needs and preferences. This requires close collaboration with a multidisciplinary team consisting of occupational therapists, physiotherapists, and speech-language pathologists to establish comprehensive and attainable goals.

An effective aquatic therapy program for a child with autism may involve exercises targeting joint attention, sensory processing, motor skills, and social interaction, all conducted in the therapeutic environment of water. Activities should be structured to provide predictability and routine, which can help alleviate anxiety in participants with ASD. Additionally, incorporating the child's interests into therapy sessions can enhance engagement and motivation. Safety is of utmost importance in aquatic therapy, considering that children with autism may have varying swimming abilities and comfort levels in water. This requires conducting individual risk assessments and implementing appropriate safety measures, including continuous supervision by trained professionals [11].

7.4 Historical Context of Aquatic Therapy

Aquatic therapy, which involves utilizing water for pain relief and treatment, has a long history spanning thousands of years. Ancient civilizations such as the Egyptians, Greeks, and Romans recognized and utilized water's healing properties. The Greeks, for example, utilized hot springs for relaxation and mental well-being, while the Romans engineered sophisticated baths for hygiene and therapeutic purposes.

In the nineteenth century, Vincent Priessnitz, a farmer from what is now the Czech Republic, is credited with modernizing aquatic therapy. He developed a system of cold-water treatment after observing animals using cold river water to heal their injuries. Priessnitz's methods gained popularity in Europe and later in America, contributing to the evolution of contemporary aquatic therapy practices [19, 20].

7.5 Principles of Aquatic Therapy

Aquatic therapy is rooted in the physical attributes of water, namely, buoyancy, resistance, hydrostatic pressure, and temperature modulation, which contribute to its therapeutic benefits. Buoyancy lessens gravitational impact on the body, reducing joint stress and facilitating exercises that may be challenging on land. Resistance assists in muscle strengthening and conditioning. Hydrostatic pressure can reduce swelling and enhance circulation. Temperature regulation, using warm or cold water, can alleviate muscle tension, pain, and promote circulation. These principles are employed across various aquatic therapy techniques, including whirlpool baths, aquatic exercises, and contrast water therapy, to address diverse conditions and promote overall well-being [21, 22].

7.6 Water's Properties and Their Therapeutic Benefits

Water provides a unique therapeutic environment due to its physical properties, including buoyancy, resistance, hydrostatic pressure, and thermal conductivity. Each of these properties plays a role in the therapeutic benefits of aquatic therapy for individuals with ASD:

- *Buoyancy:* Buoyancy reduces the effects of gravity on the body, allowing individuals with motor difficulties to move more freely and with less effort in water. This can help improve motor skills, balance, and coordination.
- *Resistance:* Water offers natural resistance in all directions, which can help strengthen muscles and improve motor planning and execution. This resistance is proportional to the effort exerted against it, allowing for customized intensity levels.
- *Hydrostatic pressure:* The pressure exerted by water increases with depth, providing proprioceptive feedback and enhancing sensory integration. This pressure can also help reduce sensory overload by offering a calming, enveloping sensation.
- *Thermal conductivity:* Warm water can help relax muscles and increase circulation, providing comfort and reducing muscle spasticity. The warmth of the water can also have a calming effect, reducing anxiety and promoting relaxation [23].

7.6.1 Psychological Benefits of Water-Based Therapies

Beyond the physical benefits, aquatic therapy offers several psychological advantages for individuals with ASD. The sensory experience of being in water can be inherently soothing, helping to reduce stress, anxiety, and behavioral issues associated with sensory dysregulation. The water environment also offers a novel, playful setting that can enhance motivation and engagement in therapy. Participating in water-based activities can improve self-esteem and social skills by providing opportunities for positive social interactions in a less structured, more forgiving environment than traditional therapy settings [24].

Aquatic therapy offers a multifaceted approach to supporting children with autism spectrum disorder (ASD), addressing physical, sensory, cognitive, emotional, and social needs in a therapeutic setting. The unique properties of water provide a supportive environment for engaging in therapeutic activities, leading to various benefits [25].

7.6.2 Physical Benefits of Aquatic Therapy: Improved Motor Skills, Balance, and Coordination

Aquatic therapy facilitates physical development in children with ASD by leveraging the buoyancy of water to reduce the effects of gravity, making it easier to perform exercises that improve motor skills, balance, and coordination. The resistance offered by water enhances muscle strength and endurance without the need for weights, making the exercises safer and more manageable for children with varying levels of physical abilities [14].

7.6.3 Sensory and Cognitive Benefits of Aquatic Therapy: Enhanced Sensory Processing, Concentration, and Focus

Water provides a unique sensory environment that can help children with ASD improve sensory processing abilities. The hydrostatic pressure and the consistent tactile stimulation from water can help regulate sensory overload, a common challenge for individuals with ASD. This regulated sensory input can lead to improvements in concentration, focus, and overall cognitive functioning as children become less distracted by sensory discomfort. [3]

7.6.4 Emotional and Social Benefits of Aquatic Therapy: Stress Reduction, Increased Self-Esteem, and Better Social Interactions

The soothing nature of water, combined with the enjoyable aspects of aquatic activities, can significantly reduce stress and anxiety levels in children with

ASD. Participating in group water therapy sessions also offers opportunities for social interaction in a noncompetitive, relaxed environment, which can enhance social skills and encourage friendships. Successes achieved during therapy can boost self-esteem and confidence, contributing to a positive self-image [16].

7.7 Case Studies and Research Findings Supporting Aquatic Therapy Benefits

Case studies and research findings provide empirical support for the benefits of aquatic therapy in children with ASD. For instance, a case study of a child with ASD participating in a tailored aquatic therapy program reported significant improvements in social interaction and communication skills and a reduction in stereotypical behaviors after participating in regular sessions over several months [16].

Implementing aquatic therapy for children with ASD involves a careful and thoughtful process to ensure that each child receives the most benefit from their sessions. This process includes assessment, program design, goal setting, personalization, activity selection, and safety considerations [16].

7.8 Assessing a Child's Suitability for Aquatic Therapy

Before beginning aquatic therapy, a thorough assessment by a team of professionals, including occupational therapists, physiotherapists, and specialists in ASD, is crucial. This assessment should evaluate the child's physical abilities, sensory sensitivities, medical history, and any contraindications to water-based activities. Understanding the child's current level of functioning across different domains ensures that aquatic therapy is a suitable and safe option [26].

7.9 Designing Individualized Aquatic Therapy Programs

Based on the initial assessment, a customized aquatic therapy program is designed to address the specific needs and goals of the child. This program should consider the child's strengths, areas for improvement, and any particular interests that could enhance their engagement in therapy [26].

7.9.1 Setting Goals Based on the Child's Needs and Abilities

Goal setting is a collaborative process involving therapists, the child, and their caregivers. Objectives should be SMART (specific, measurable, achievable, relevant, time-bound), focusing on areas such as improving motor skills, enhancing sensory processing, reducing anxiety, or fostering social skills. Goals should be revisited and adjusted as needed to reflect the child's progress and evolving needs [13].

7.9.2 Incorporating Interests and Motivational Factors

Incorporating a child's interests into the aquatic therapy sessions can significantly increase motivation and participation. Activities that align with the child's preferences, whether it is related to certain toys, themes, or types of play, can make therapy more enjoyable and engaging [16].

7.9.3 Key Exercises and Activities Tailored for Autistic Children

Exercises and activities should be chosen to address the child's specific therapeutic goals. For example, activities that involve reaching for and grasping objects can improve fine motor skills and hand-eye coordination, while floating and moving through water can enhance body awareness and balance [27].

7.9.4 Examples of Activities Targeting Specific Therapeutic Goals

- For motor skills: Walking or swimming against the pool's resistance to strengthen muscles
- For sensory processing: Playing with water toys that have different textures or produce different sounds to provide varied sensory inputs
- For social skills: Participating in group games that require taking turns and following instructions

7.9.5 Safety Considerations and Best Practices

Safety is of utmost importance in aquatic therapy. This involves ensuring that the water temperature is safe, maintaining clear communication among all involved parties, and having trained professionals such as lifeguards present during sessions. All equipment used should undergo safety checks, and emergency procedures should be established and understood by staff.

Implementing aquatic therapy for children with autism spectrum disorder (ASD) presents unique challenges and considerations. Addressing these effectively requires an understanding of the specific needs of children with ASD, a comprehensive approach to safety, and strategies tailored to engage and motivate participants of various ages and abilities [15].

7.10 Addressing Common Challenges in Aquatic Therapy for Autistic Children

One of the primary challenges is the varied response to sensory stimuli among children with ASD. Some may find the sensation of water overwhelming, while others may be under-responsive and unaware of the dangers water can pose. Additionally, communication difficulties can make it hard for therapists to understand a child's discomfort or enjoyment within the session.

Strategies: [14]

- Gradual acclimatization to water to reduce sensory overload
- Use of visual supports and non-verbal communication cues to enhance understanding
- Consistent routines and familiar equipment to provide a sense of security

7.11 Mitigating Risks and Ensuring Safety

Safety is a significant concern in aquatic therapy sessions. Children with ASD might not recognize dangerous situations, such as being too close to the pool's edge or not understanding the depth of water.

Strategies: [7]

- Always have trained lifeguards or therapists with aquatic rescue training present.
- Use appropriate floatation devices for nonswimmers.
- Maintain a low therapist-to-child ratio to ensure close supervision.

7.12 Considerations for Different Ages and Abilities

Children with ASD present a wide range of abilities and developmental stages, necessitating a flexible approach to aquatic therapy.

Strategies:

- Tailor activities to match the child's developmental level and physical capabilities.
- Adjust the depth of water and temperature based on the child's age, size, and sensory sensitivities.
- Incorporate adaptive equipment for children with motor impairments.

7.13 Strategies for Engaging and Motivating Children

Keeping children with ASD engaged and motivated in aquatic therapy can be challenging, particularly for those with limited attention spans or difficulty understanding the purpose of activities.

Strategies:

- Incorporate the child's interests into therapy sessions to increase engagement.
- Use positive reinforcement and celebrate achievements, no matter how small.
- Introduce variety in activities while maintaining a predictable structure to sessions [28].

7.14 Discussion

The incorporation of aquatic therapy into treatment plans for children with ASD has gained significant interest in recent years. This review examines the current state of research, identifies areas where research is lacking, and examines the viewpoints of therapists and parents regarding the efficacy of aquatic therapy. Ultimately, it provides a summary of important findings and considerations for the future of aquatic therapy in autism care.

7.15 Analysis of Current Research on Aquatic Therapy and Autism

Recent studies have emphasized the diverse benefits of aquatic therapy for children with ASD, showing improvements across physical, sensory, emotional, and social aspects. For instance, Yilmaz et al. [10] found enhancements in physical fitness and water orientation after swimming training for children with autism. Similarly, researchers observed notable social and behavioral improvements following aquatic exercise programs.

However, research in this field often encounters methodological challenges, such as small sample sizes, absence of control groups, and reliance on subjective reporting measures, which can impact the generalizability and conclusiveness of findings.

7.15.1 Gaps in the Literature and Areas for Future Research

One significant research gap is the lack of long-term longitudinal studies that track the effects of aquatic therapy on ASD symptoms and quality of life over time. Additionally, more rigorous and controlled studies are needed to better comprehend how aquatic therapy produces its benefits and to determine the most effective treatment protocols. Future research should also investigate how integrating aquatic therapy with other therapeutic approaches can optimize outcomes for children with ASD.

Therapists and Parents Perspectives on the Effectiveness of Aquatic Therapy
Both therapists and parents frequently share positive feedback about aquatic therapy, noting enhancements in children's physical abilities, social skills, and overall behavior. They attribute these improvements to the sensory-rich and low-stress water environment. However, challenges such as limited facility access, cost, and the necessity for specialized therapist training are identified as obstacles to broader adoption.

7.16 Conclusion

Summary of Key Points
The Potential of Aquatic Therapy as Part of a Holistic Approach to Autism Care

- Aquatic therapy provides a distinct and valuable therapeutic approach for children with ASD, targeting physical, sensory, emotional, and social needs.
- Ongoing research shows promise but is hindered by methodological limitations, highlighting the necessity for more rigorous studies.
- Therapists and parents alike observe favorable results with aquatic therapy, despite encountering obstacles related to access and affordability.

Aquatic therapy has significant potential as part of a comprehensive care approach for children with ASD, serving as a complementary method that can augment the effectiveness of conventional therapies. Its capacity to create a customized, sensory-friendly setting makes it well-suited for addressing the intricate needs of individuals with autism spectrum disorder.

Final Thoughts on Advancing Aquatic Therapy Practices for Autistic Children
Improving aquatic therapy practices for children with ASD necessitates addressing current obstacles such as enhancing facility access, reducing expenses, and improving therapist training. Investing in robust research to strengthen the evidence supporting aquatic therapy's effectiveness is also vital. As knowledge and resources expand, aquatic therapy has the potential to become a more essential component of autism care, leading to enhanced outcomes and quality of life for affected children and their families.

References

1. Vonder Hulls DS, Walker LK, Powell JM. Clinicians' perceptions of the benefits of aquatic therapy for young children with autism: a preliminary study. Phys Occup Ther Pediatr. 2006;26(1–2):13–22.
2. Cole AJ. Comprehensive aquatic therapy. Butterworth Heinemann. 2004;11(1):123–9.
3. Prior M. Is there an increase in the prevalence of autism spectrum disorders? J Paediatr Child Health. 2003;39(2):81–2.
4. Branjerdporn G, Meredith P, Strong J, Garcia J, Brockington IF, Aucamp HM, et al. American Psychiatric Association. Diagnostic and statistical manual of mental disorders, 5th ed. Depression. 2013;16(4):813–31.
5. Ayres AJ. Sensory integration and learning disorders. Pediatr Health. 1972;10(6):221–8.
6. Brondino N, Fusar-Poli L, Rocchetti M, Provenzani U, Barale F, Politi P. Complementary and alternative therapies for autism spectrum disorder. Evid Based Complement Alternat Med. 2015;2015:258289.
7. Pan CY. Effects of water exercise swimming program on aquatic skills and social behaviors in children with autism spectrum disorders. Autism. 2009;14(1):9–28.
8. Frye RE, Rose S, Boles RG, Rossignol DA. A personalized approach to evaluating and treating autism spectrum disorder. J Pers Med. 2022;12(2):147–54.
9. Lane SJ, Mailloux Z, Schoen S, Bundy A, May-Benson TA, Parham LD, Smith Roley S, Schaaf RC. Neural foundations of Ayres sensory integration. Brain Sci. 2019;9(7):153–9.
10. Morton J, Montgomery P. Youth empowerment program for autism spectrum disorder: using paediatric aquatic therapy. J Paediatr Child Health. 2013;49(1):E19–22.
11. Fragala-Pinkham M, Haley SM, O'Neil ME. Group aquatic aerobic exercise for children with disabilities. Disabil Rehabil. 2010;32(20):1635–43.
12. Larsen J. Aquatic therapy: what is it and how can it help your child? Peads Care. 2020;24(11):221–9.
13. Güeita-Rodríguez J, Ogonowska-Slodownik A, Morgulec-Adamowicz N, Martín-Prades ML, Cuenca-Zaldívar JN, Palacios-Ceña D. Effects of aquatic therapy for children with autism spectrum disorder on social competence and quality of life: a mixed methods study. Int J Environ Res Public Health. 2021;18(6):3126.
14. Yilmaz I, Yanarda M, Birkan B, Bumin G. Effects of swimming training on physical fitness and water orientation in autism. Pediatr Int. 2004;46(5):624–6.
15. Case-Smith J, Arbesman M. Evidence-based review of interventions for autism used in or of relevance to occupational therapy. Am J Occup Ther. 2008;62(4):416–9.
16. Mortimer R, Privopoulos M, Kumar S. The effectiveness of hydrotherapy in the treatment of social and behavioral aspects of children with autism spectrum disorders: a systematic review. J Multidiscip Healthc. 2014;7:93–104.
17. Schaff RC, Smith Roley S. Sensory integration: applying clinical reasoning to practice with diverse populations. Pro-Ed. 2006;18(1):229–303.
18. Kinnaird DW, Becker BE. Contemporary aquatic therapy and pain management. In: Integrative pain medicine: the science and practice of complementary and alternative medicine in pain management. 2008. p. 285–306.
19. Moffat M. The history of physical therapy practice in the United States. J Phys Ther Educ. 2003;17(3):15–25.
20. Gianfaldoni S, Tchernev G, Wollina U, Roccia MG, Fioranelli M, Gianfaldoni R, Lotti T. History of the baths and thermal medicine. Maced J Med Sci. 2017;5(4):566–8.
21. Edlich RF, Towler MA, Goitz RJ, Wilder RP, Buschbacher LP, Morgan RF, et al. Bioengineering principles of aquatic therapy. J Burn Care Rehabil. 1987;8(6):580.
22. Lord C, Bishop SL. Autism spectrum disorders: diagnosis, prevalence, and services for children and families. Soc Policy Rep. 2010;24(2):1–2.

23. Chu CH, Pan CY. The effect of peer- and sibling-assisted aquatic program on interaction behaviors and aquatic skills of children with autism spectrum disorders and their peers/siblings. Res Autism Spectr Disord. 2012;6(3):1211–23.
24. Mooventhan A, Nivethitha L. Scientific evidence-based effects of aquatic therapy on various systems of the body. N Am J Med Sci. 2014;6(5):199–209.
25. Getz M, Hutzler Y, Vermeer A. Effects of aquatic interventions in children with neuromotor impairments: a systematic review of the literature. Clin Rehabil. 2006;20(11):927–36.
26. Oriel KN, Kanupka JW, George CL, Himmelberger B, Janke B, Repoley M. The impact of participation in a structured aquatic exercise program on parents' perceptions of behavior in children with autism spectrum disorder. J Aquat Phys Ther. 2017;25(1):13–21.
27. Sefen JAN, Al-Salmi S, Shaikh Z, AlMulhem JT, Rajab E, Fredericks S. Beneficial use and potential effectiveness of physical activity in managing autism spectrum disorder. Front Behav Neurosci. 2020;14:587560.
28. Uljarević M, Frazier TW, Jo B, et al. Relationship between social motivation in children with autism spectrum disorder and their parents. Front Neurosci. 2021;15:660330.

Applied Behavioral Analysis (ABA): A Guide to Autism

8

Riya Kalra, Kanu Goyal, and Manu Goyal

8.1 Introduction

Applied behavioral analysis (ABA) is the approach behind treatment which include behavior analysis. ABA is based on behaviorism, education of behavior, and application science, and its methodologies may be used to implement a number of intervention strategies for children having. While application-based research is increasing, meta-analyses indicate that additional study is needed in order to fully comprehend the effectiveness and efficacy of ABA interventions. Providing more evidence can help clear up misconceptions and reduce overuse of ABA therapies [1].

Applied behavior analysis (ABA), an effective treatment for ASD that employs the concepts of behavioral theory, theory of learning, and reinforcement strategies to achieve behavioral goals, is widely regarded as the gold standard, scientifically proven treatment for ASD that may enhance quality of life, behavior, and communication [2].

ABA is a strategy that uses psychological ideas from the theory of learning to modify repetitive behaviors in persons with ASD. In the Lovaas devised an approach based on B. F. Skinner's operant conditioning theory to help treat children with ASD by modifying their behaviors to improve social connections. "California University, Young Autism Research model" was established and scientifically verified to examine the beneficial effects of one-on-one intervention for children with ASD over a 2–3-year period [3].

Over the past 60 years, ABA has grown from the basic principles established in the previously Lovaas model and then the UCLA Young Autism Project to much more comprehensive approaches to treatment centered on intervention procedures, processes, and methods of instructions, all of which aim to address deficiencies for

R. Kalra · K. Goyal (✉) · M. Goyal
Maharishi Markandeshwar Institute of Physiotherapy and Rehabilitation, Maharishi Markandeshwar (Deemed to be University), Mullana - Ambala, Haryana, India
e-mail: kanu.goyal@mmumullana.org; manu.goyal@mmumullana.org

M. Goyal, K. Goyal (eds.), *Rehabilitation Approach in Autism*,
https://doi.org/10.1007/978-981-96-4162-8_8

children with ASD at all phases of performance, which includes thinking and conversing, social skills, issues with behavior, and ability to live independently [4].

8.2 ABA-Based Treatment Models

The most widely used ABA-based therapy programs include early intensive behavior intervention (EIBI), the Early Starting Denver Model (ESDM), and Learning Experiences and Alternative Program for kids in preschool and their parents (LEAP) [5].

EIBI is geared for children under the age of 5, and it is often delivered for 20–40 h per week over several years. Discrete trial training (DTT) or naturalistic training is frequently employed in individual sessions in structured environments such as homes or schools. This holistic therapeutic paradigm addresses all aspects of functioning, including living independently, social interactions, motors, and educational ability, as well as language [6].

The ESDM was established for kids with ASD aged between the ages of 12 and 60. This intervention builds on ABA's naturalistic teaching methodologies to provide a comprehensive, developmental, and relationship-based behavioral modification for children in their infancy. Recently, numerous complete ABA therapy models have transitioned from severe, operational conditioning-based one-on-one models to more practical and generalizable programming [7].

LEAP is a model for children with autism that is implemented in public schools. LEAP depends on ABA principles and uses a variety of methods, including pivotal response teaching (PRT), timing delay, incidental teaching, peer-mediated measures, and PECS. It is significant which the fundamental concept of LEAP emphasizes parental and peer participation when instructing behavioral techniques and relies on organically associated teaching arrangements, as opposed to the directed, adult-driven guidance used in a majority of the other separated ABA intervention strategies [8].

8.3 Definition of ABA

Definitions range from strict rules for some ABA-based programs to collections of specific ABA techniques, in addition to ABA as a structure for assessing practices instead of a measure in and of itself [9].

Granpeesheh et al. defined ABA as "the implementation of educational and inspirational principles to find solutions to socially significant challenges." The notion of ABA as an investigation approach is comparable to the idea of Baer et al. from the late twentieth century. It refers to psychological studies that are utilized, behavioral, analytic, technical, theoretically structured, and successful, with the possibility for generalized consequences [10].

8.4 ABA in Autism Spectrum Disorder (ASD)

Autism spectrum disorders (ASDs) include autistic disorder, chronic developmental condition not otherwise defined, Asperger's syndrome, Rhett's sickness, and childhood disintegrative disorder (299.10). ASD is distinguished by deficiencies in two or more areas of functioning, including delayed language growth, delayed social development, or a high level of excessively and stereotypically repetitive interests or hobbies. ASD statistics have grown significantly in recent years, which might be related to both a rise in the incidence of the disease and more understanding of the problem among clinicians [11].

Several independent organizations, including the US Physician General, the New York City State Department for Wellness, the National Academy of Sciences, and the American Board of Pediatrics, have recognized and recommended the application of ABA among kids with autism, citing a considerable body of evidence [12].

The impact of such support may be observed in policy developments such as formal state funding for autism treatment and statewide-level laws requiring health coverage for ABA treatment [13].

ABA is the science of examining whether an individual's environment influences their behavior and detailing methods that employ the findings to enhance behavior. It is theoretically based on operant conditioning and attempts to examine and correct problematic behavior while encouraging and generalizing more adaptive actions, such as via systematic reinforcement [14].

According to 2002 data from the Centers for Disease Control and Prevention, the autism prevalence among children in the United States is around one out of every 150. As a result, ASD has garnered greater public interest in the past few years, as well as substantial funding for research. However, parent groups typically lament the lack of research and meaningful data on effective therapy.

There are various treatments for ASD which are generally endorsed by experts and parents looking for a cure. Unfortunately, the majority of these treatments lack scientific support. ABA-based intervention is a widely accepted therapeutic method. ABA was originally used for the treatment of autism in the beginning of the 1960s and continues to be thoroughly researched in peer-reviewed journals.

ABA-based therapies can target particular actions (e.g., toilet training) while improving cognition. ABA-based therapies begin in early childhood, often between 3 and 4 years old; are intensive (20–40 h per week), individualized to fit individual requirements; cover numerous skills all at once (e.g., joint attention); and include a variety of behavior analytic methodologies. In addition, comprehensive ABA-based treatment include one-on-one sessions supplemented by actions for groups and transgender individuals.

Intensive ABA intervention can enhance adaptive behavior, reduce ASD severity, and boost academic achievement [15, 16].

8.5 Principles of ABA

The core learning principle of ABA is the idea that an action can either strengthen or weaken its effects. An action that is followed by the provision of beneficial results or the elimination of aversive consequences will be strengthened (referred to as reinforcement), yet an action that is followed by the emergence of negative implications or the removal of desired outcomes will be weakened [17].

We can influence the environment to either increase or decrease behavior in the future by consistently enforcing such consequences when it occurs. Extinction is the idea that stopping receiving an incentive for a behavior reduces the likelihood that the action will occur again.

Reinforcers conditioned to elicit the same rise in behavior as the original reinforcer when they coexist with additional stimuli. This strategy is known as conditional reinforcement. Moreover, an action starts to occur only in the presence of its antecedent stimulus and not in its absence when it reinforces itself when it occurs in the surroundings of that stimulus. "This method is known as stimuli control." When you stop on a red light while driving, you avoid getting into an accident and getting a ticket. This is due to a past in which that behavior was reinforced in the presence of a certain stimulus, which now indicates which actions will lead to which reinforcers [18].

Finally, the capacity to transfer acquired behaviors to different settings, enabling the acquisition of abilities in one, is the result of rewarding an action when evidence of stimuli transference in additional stimuli is present. The act of translating the knowledge gained from one situation to another is known as generalization. It includes all of the elements that make up the historical context in which an event occurs, such as changes to the setting, the people there, and the moment of day when the action occurs [19].

8.6 Implication of Principles of ABA

The concepts of reward, extinction, sensory control, and generalization are used in four teaching procedures: prompting, fading, shaping, and chaining.

Using cues or assistance to elicit an action that would not otherwise happen is known as prompting. For instance, in order to educate an autistic child to ask for a certain automobile, one could use the instruction "Say car." This would encourage the youngster to say "car," and the therapist could then reward the behavior by giving the child his favorite car. Prompts are thought of as a transient instrument for inducing and supporting particular actions [20].

Fade must be applied almost immediately. The practice of methodically eliminating a cue such that the intended behavior or capacity can continue even in its absence is known as fading. Depending on how serious and immediate they are, prompts may be dimmed [21].

A method for promoting incremental estimations toward a desired result is called shaping. If a toddler cannot replicate the complete word "car," for instance, you might start by encouraging them to copy only the pronunciation of "K." After the child has mastered this skill, one can go on to saying and repeating "kah" and then "car."

A technique for teaching lengthy sequences of behaviors that could be challenging for kids with ASD to understand all at once is called chaining. A "chain" of behaviors is created by breaking down the sequence into smaller acts that are prompted and reinforced. Identifying the essential ingredients, taking them out of the refrigerator, setting them out on the counter, preparing the snack, cleaning up, and repeating the process until every step is learned are the first steps in teaching a kid how to make a snack [22].

It is necessary to incorporate the concepts and methods covered above into thorough teaching forms. Since a discrete trial (DTT) was a major component of early EIBI programs, it is possible that this has led to some confusion between DTT and EIBI, even though the two are different instructional methods. Throughout the day, several discrete opportunities, or "discrete trials," are presented as part of the systematic teaching strategy known as DTT. In an isolated trial, the child reacts to an instruction from the therapist, and the therapist then imposes a consequence [23].

For example, the therapist could give a child a reinforcer when they answer questions like "What is your surname?" well. When a child replies incorrectly, the therapist could choose to start a fresh discrete trial and mimic the appropriate reaction as a remedial measure rather than offering a reinforcer. The strength of DTT is in its capacity to offer a large number of repetitions in a little length of time, offering several learning opportunities [24].

One drawback of DTT is that certain children could find the structured and exacting training methods unpleasant, and the skills acquired might not be useful in less structured situations. Even though preliminary research has validated these concerns, more investigation is necessary. As will be discussed below, modern EIBI programs usually supplement DTT with less structured teaching techniques [25].

NET seeks to replicate typical adult-child interactions and maximize learning opportunities, in contrast to DTT's highly structured and artificial learning environment. NET uses an environment that is analogous to a young child's everyday activities to teach skills. In contrast to DTT, NET allows learners to initiate learning attempts instead of the therapist. When a child grasps, points, or vocalizes something they want, the learning trial starts. The therapist recognizes a teaching moment and initiates a desired behavior before giving the child what they require [26].

NET includes a number of different approaches, such as milieu teaching, incidental education, and critical response training. Improved skill acquisition and a less structured style are two advantages of NET that may lessen adverse responses in certain children. While DTT and NET form the foundation of fully functional EIBI programs, other ABA techniques have also been carefully studied and extensively applied in EIBI programs. While not every approach is covered in length on this page, readers may learn more about script-fading, visual activity calendars, peer-mediated social abilities training, and video modeling [27].

8.7 Challenging Behavior

Repetitive behavior is an intrinsic component of ASD diagnostics. Conversely, other maladaptive behaviors, including as aggression, self-harm, tantrums, and property destruction, are also rather common and often result from delays in communication skills. Because all unacceptable actions have an opportunity to impede learning and day-to-day functioning, effective EIBI programs must properly address them. When addressing these types of behaviors, the ABA approach examines the reasons behind the actions and searches for the recurrent sources of reinforcing that they usually lead to. Unintentional encouragement from caregivers can take many forms, such as providing attention, letting go of unpleasant chores, and granting access to preferred hobbies, foods, or products. Unable to communicate verbally, a child may regularly act out or have tantrums to try to achieve what they want. There is a constant source of reinforcement for any bad conduct [28].

8.8 Scope of Intervention

The EIBI's programs are comprehensive. Put another way, the programs precisely address every observable area of shortcoming. To put it simply, if a comprehensive EIBI program determines that a skill is a taught behavior and that a kid should be able to perform it at their age, it should be taught to them. Among the broad curriculum areas that are taught are language, relationships, play skills, physical ability, pre-academic and educational abilities, and self-sustaining living skills [29].

8.9 Weekly Frequency of Therapeutic Intensity

The number of hours of the EIBI must be done each week is a topic of recurrent disagreement. But the investigation's conclusions are rather clear-cut. In Lovaas16's initial experiment, a group received a low-in ABA (approximate 10 h per week); the results revealed much smaller rises compared to those of the intense (40 h each week) group. For 2 years, or around 12 h a week, autistic children participated in low-intensity ABA therapy, according to a study by Eldevik et al. These individuals' advantages were statistically significant in comparison to a control group, although they were much less than those shown in high-intensity ABA studies [30].

In a descriptive research, Reed and associates looked at the 9-month outcomes of two child groups receiving ABA therapy: One group received 30 h on average per week, while the other received 12 h on average. They found that the children in the intense group advanced noticeably faster. Smith et al. evaluated the effects of 25 h per week of ABA therapy for 2 years. The results indicated that the ABA group's benefits were statistically significant when compared to a control group, but significantly smaller than those observed in high-intensity ABA studies. Combining the research demonstrates that providing EIBI alone is not enough to provide the optimal outcomes; rather, EIBI has to be sufficiently intense [31].

8.10 Total Duration of Treatment

The amount of period (years, months, etc.) which an autistic child needs for therapy so as to make the greatest development is another topic of widespread controversy. However, some studies suggest that therapy should ideally persist for a minimum of 2 years to have the optimum results. Howard and colleagues compared intensive ABA therapy for 14 months versus eclectic therapy over a 14-month period for young children with autism. In regards to statistically significant gains, the ABA group did better than the control group; however, the advantages were not as great as those seen in studies assessing intensive ABA for longer periods of time [32].

Although the follow-up data from the Howard et al. study has not yet been released, it shows that the difference in outcomes between the two EIBI group and the control group increased throughout the course of the next year of EIBI exposure. Even with significant improvements shown following a year of intensive ABA therapy, the results of research by Eikeseth and colleagues were not as high as those of longer-term ABA therapy. Additional data from the same study was later published, and it showed that children in the behavioral treatment group continued to make significant progress in the second year of therapy [33].

Studies looking at long-term ABA revealed larger but less significant benefits than did vigorous ABA for 9 months, as reported by Reed et al. Results on outcomes following the first year for treatment (rather than the complete term, which equals 4 years) were reported by Sallows and Graupner, who found that children undergoing EIBI exhibited considerable progress after 1 year of therapy when compared to controls. However, throughout the following 3 years, participants kept improving [34].

Children receiving less intensive and shorter-term ABA therapy had significant increases in IQ when compared to controls; however, these enhancements were not as large as those shown in studies that employed ABA therapy over a 2-year period. Zachor et al. found that although all of the children in the developmental group still met the requirements for an ASD diagnosis, the kids receiving intensive ABA were no longer eligible for one. This was after comparing the outcomes of autism-related interventions, such as growth therapy and intensive ABA, for a year. Although the results are compelling, other studies that examined the use of EIBI over longer time periods found that a higher percentage of people rated in the non-impaired [35].

8.11 Determinants of the Course of Therapy

Even though EIBI has been found to greatly enhance the lives of children with ASD, the intensity of response to treatment appears to differ significantly throughout children. The necessity for research on factors that predict result has been highlighted by the repeated finding of EIBI outcomes studies. The characteristics of children who gain the most from EIBI have been identified via preliminary research. For example, Bono et al. found that improvements achieved by participants following the intervention were correlated with their initial language skills and their ability to respond to demands for shared attention from others. Sigman and McGovern40 discovered that the frequency of inquiring behaviors and functional play abilities were indicative of treatment effectiveness [36].

Sallows and Graupner discovered a link between the pretreatment skills of language, sociability, and imitation and the outcomes of therapy. Additionally, nonverbal cognitive capacity and early development of language were found to be predictive of outcome by Szatmari and colleagues. Although these studies are beginning to examine the connection between a child's distinctive characteristics and how they react to intervention, the variety of their results also highlights how challenging it is now to predict which children will benefit most from EIBI [37].

8.12 Summary

Applied behavior analysis (ABA) is a scientifically supported treatment for ASD that utilizes behavioral theory and reinforcement strategies to improve communication, social skills, and overall quality of life. Grounded in operant conditioning, ABA methods aim to modify behaviors through structured interventions like Early Intensive Behavior Intervention (EIBI), the Early Starting Denver Model (ESDM), and the Learning Experiences Approach (LEAP).

EIBI, often delivered for 20–40 h a week, focuses on individual skills in structured settings, while ESDM incorporates naturalistic teaching methods for younger children. LEAP emphasizes peer and parental involvement within school environments.

Research indicates that intensive ABA programs yield significant improvements, particularly when applied for at least 2 years. However, outcomes can vary based on factors like initial language skills and engagement levels. Effective ABA intervention not only addresses challenging behaviors but also promotes skill acquisition across various domains, enhancing independence and adaptive functioning. Despite its recognition and endorsements from major health organizations, ongoing studies are necessary to further validate ABA's effectiveness and address misconceptions surrounding its use.

8.13 Conclusion

The intensity and duration of ABA therapy are critical for achieving significant outcomes, with studies indicating that higher intensity and longer treatment periods yield better results. However, individual factors such as initial language skills and cognitive abilities significantly influence treatment responsiveness, highlighting the complexity of predicting outcomes. Overall, ABA remains a cornerstone in ASD intervention, supported by a substantial body of evidence and formal endorsements from various health organizations. Further research is essential to refine approaches and optimize therapeutic benefits.

References

1. Foxx RM. Applied behavior analysis treatment of autism: the state of the art. Child Adolesc Psychiatr Clin N Am. 2008;17(4):821–34.
2. Fisher WW, Piazza CC, Roane HS, editors. Handbook of applied behavior analysis. Guilford Publications; 2021.
3. Bailey JS, Burch MR. Research methods in applied behavior analysis. Routledge; 2017.
4. Alberto PA, Troutman AC. Applied behavior analysis for teachers. 5th ed. Merrill/Prentice-Hall: Upper Saddle River; 1999.
5. Wahler RG, Fox JJ. Setting events in applied behavior analysis: toward a conceptual and methodological expansion. J Appl Behav Anal. 1981;14(3):327–38.
6. Roane HS, Fisher WW, Carr JE. Applied behavior analysis as treatment for autism spectrum disorder. J Pediatr. 2016;175:27–32.
7. Critchfield TS, Reed DD. The fuzzy concept of applied behavior analysis research. Behav Anal. 2017;40(1):123–59.
8. Peterson T, Dodson J, Strale F Jr. Impact of applied behavior analysis on autistic children target behaviors: a replication using repeated measures. Cureus. 2024;16(2):221–9.
9. Alves FJ, De Carvalho EA, Aguilar J, De Brito LL, Bastos GS. Applied behavior analysis for the treatment of autism: a systematic review of assistive technologies. IEEE Access. 2020;8:118664–72.
10. Choi KR, Bhakta B, Knight EA, Becerra-Culqui TA, Gahre TL, Zima B, et al. Patient outcomes after applied behavior analysis for Autism-Spectrum-Disorder. J Dev Behav Pediatr. 2022;43(1):9–16.
11. Eckes T, Buhlmann U, Holling HD, Möllmann A. Comprehensive ABA-based interventions in the treatment of children with autism spectrum disorder–a meta-analysis. BMC Psychiatry. 2023;23(1):133–41.
12. Matson JL. Applied behavior analysis for children with autism spectrum disorders. New York: Springer; 2009.
13. Cebula KR. Applied behavior analysis programs for autism: sibling psychosocial adjustment during and following intervention use. J Autism Dev Disord. 2012;42(5):847–62.
14. Kelly MP, Alireza I, Busch HE, Northrop S, Al-Attrash M, Ainsleigh S, et al. An overview of autism and applied behavior analysis in the Gulf Cooperation Council in the Middle East. Rev J Autism Dev Disorders. 2016;3(2):154–64.
15. Larsson EV. Is applied behavior analysis (ABA) and early intensive behavioral intervention (EIBI) an effective treatment for autism? A cumulative history of impartial independent reviews. Autism. 2013;27(1):168–792.
16. Leaf JB, Leaf R, McEachin J, Taubman M, Ala'i-Rosales S, Ross RK, et al. Applied behavior analysis is a science and, therefore, progressive. J Autism Dev Disord. 2016;46(2):720–31.
17. Fisher WW, Luczynski KC, Blowers AP, Vosters ME, Pisman MD, Craig AR, et al. A randomized clinical trial of a virtual-training program for teaching applied-behavior-analysis skills to parents of children with autism spectrum disorder. J Appl Behav Anal. 2020;53(4):1856–75.
18. Fortunato JA, Sigafoos J, Morsillo-Searls LM. A communication plan for autism and its applied behavior analysis treatment: a framing strategy. In: Child and youth care forum 2007, vol 36. Kluwer Academic Publishers-Plenum Publishers; 2007 p. 87–97.
19. Chang K, Zaroff CM. Applied behavior analysis in autism spectrum disorders in China and Hong Kong. Acta Psychopathol. 2017;3(5):11.
20. Choutka CM, Doloughty PT, Zirkel PA. The "discrete trials" of applied behavior analysis for children with autism: outcome-related factors in the case law. J Spec Educ. 2004;38(2):95–103.
21. Leblanc L, Richardson W, McIntosh J. The use of applied behavioral analysis in teaching children with autism. Int J Special Educ. 2005;20(1):13–34.
22. Virués-Ortega J. Applied behavior analytic intervention for autism in early childhood: meta-analysis, meta-regression and dose–response meta-analysis of multiple outcomes. Clin Psychol Rev. 2010;30(4):387–99.

23. Makrygianni MK, Gena A, Katoudi S, Galanis P. The effectiveness of applied behavior analytic interventions for children with Autism-Spectrum-Disorder: a meta-analytic study. Res Autism Spectr Disord. 2018;51(2):18–31.
24. Wolfe PS, Condo B, Hardaway E. Socio-sexuality education for persons with autism spectrum disorders using principles of applied behavior analysis. Teach Except Child. 2009;42(1):50–61.
25. Steege MW, Mace FC, Perry L, Longenecker H. Applied behavior analysis: beyond discrete trial teaching. Psychol Sch. 2007;44(1):91–9.
26. Stanislaw H, Howard J, Martin C. Helping parents choose treatments for young children with autism: a comparison of applied behavior analysis and eclectic treatments. J Am Assoc Nurse Pract. 2020;32(8):571–8.
27. Zachor DA, Ben-Itzchak E, Rabinovich AL, Lahat E. Change in autism core symptoms with intervention. Res Autism Spectr Disord. 2007;1(4):304–17.
28. Anderson A, Carr M. Applied behaviour analysis for autism: evidence, issues, and implementation barriers. Curr Dev Disord Rep. 2021;2(11):1.
29. Griffith GM, Fletcher R, Hastings RP. A national UK census of Applied Behavior Analysis school provision for children with autism. Res Autism Spectr Disord. 2012;6(2):798–805.
30. McClannahan LE, MacDuff GS, Krantz PJ. Behavior analysis and intervention for adults with autism. Behav Modif. 2002;26(1):9–26.
31. Schoen AA. What potential does the applied behavior analysis approach have for the treatment of children and youth with autism? J Instr Psychol. 2003;30(2):125–31.
32. Abid N, Aslam S, Azeem A, Shahidi-Hamedani S. Unlocking social growth: the impact of applied behavior analysis on children with autism spectrum disorder. Int Electronic J Element Educ. 2024;16(4):509–20.
33. Almutlaq H. Saudi special educators' perceptions of applied behavior analysis for students with autism. World J Educ. 2021;11(4):18–30.
34. Ivy JW, Schreck KA. The efficacy of ABA for individuals with autism across the lifespan. Curr Dev Disord Rep. 2016;3:57–66.
35. Heath CD, McDaniel T, Venkateswara H, Panchanathan S. Improving communication skills of children with autism through support of applied behavioral analysis treatments using multimedia computing: a survey. Univ Access Inf Soc. 2021;20(1):13–30.
36. Wolfe P, Neisworth JT. Autism and applied behavior analysis. Exceptionality. 2005;13(1):1–2.
37. Smith T, Iadarola S. Evidence base update for autism spectrum disorder. J Clin Child Adolesc Psychol. 2015;44(6):897–922.

Transforming Autism Care: The Role of Play Therapy in Enhancing Development and Well-Being

9

Ankita Sharma and Moattar Raza Rizvi

9.1 Autism Spectrum Disorders (ASD) and Play Therapy: An Overview

Autism spectrum disorders are a group of complex and heterogeneous neurodevelopmental disorders with difficulties in social communication and interaction, as well as restricted, repetitive patterns of behavior, interests, or activities. The spectrum nature of ASD reflects the wide variability in presentation, severity, and associated features of any individual having the condition. ASD is strictly considered a lifelong condition: Manifestations typically become evident before the age of 3 years. However, the symptoms of ASD could be demonstrated in a wide variety among individuals; some demonstrate clear signs of it even during infancy, while others do not get diagnosed until later during their childhood or even into adulthood [1].

It is, therefore very much recommended that ASD needs to be recognized and treated early because studies show that the starting of intensive interventions at an early age showed profound cognitive enhancement, language development, and adaptive skills as compared to those not exposed [2]. Many kinds of intervention are present for ASD such as behavioral interventions, developmental interventions, and naturalistic developmental behavioral intervention, among which play therapy has, of late, come to be a valuable tool for the support of the development and well-being of children with ASD [3].

A. Sharma (✉)
Department of Physiotherapy, Amity Institute of Health Allied Sciences, Amity University, Noida, India
e-mail: asharma39@amity.edu

M. R. Rizvi
College of Healthcare Professions (CoHP), D.I.T University, Dehradun, Uttarakhand, India
e-mail: dean.healthcare@dituniversity.edu.in

M. Goyal, K. Goyal (eds.), *Rehabilitation Approach in Autism*,
https://doi.org/10.1007/978-981-96-4162-8_9

9.2 Definition of Autism Spectrum Disorders (ASD)

Autism spectrum disorder is a complex, neurodevelopmental disorder characterized by ongoing social communication and interaction difficulties, along with little or no verbal communication, and by repetitive behaviors, interests, or activities [4]. The term "spectrum" implies the wide range of symptoms, skills, and levels of impairment.

ASD diagnosis criterion has grown with time, and the latest definition is that made by the DSM-5. The DSM-5 actually mandates diagnosis based on ongoing challenges in social communication and interaction and varied contexts and one of these continues to be social-emotional reciprocity,nonverbal communication in social interactions, and in forming, maintaining, and understanding relationships [5].

9.3 Core Features of ASD

The two main areas of challenge for people with autism spectrum disorder are social communication and interaction as well as persistent patterns of restricted and repetitive behavior, interests, or activities. Most importantly, among the deficits in nonverbal communicative behaviors include the failure to understand body language and facial expressions, individuals with ASD are known for having several difficulties relating to socio-emotional reciprocity, meaning challenges initiating social contact and difficulties expressing emotions [6]. They often struggle with friendship maintenance and changes in new social environments. Other common stereotypic behaviors include hand flapping, rigid adherence to routines, and excessive preoccupation with significantly narrowed areas of interest. An almost universally observed feature is problems with sensory sensitivity, which involve either over-responsiveness or under-responsiveness to environmental stimuli. All of these core symptoms severely affect the daily life and interactions with others.

9.4 Prevalence and Etiology

There has been a notable increase in the incidence of ASD in the last few decades. With estimates now showing that approximately 1 in 54 children are diagnosed with the condition within the United States, the reasons for such a trend are not entirely clear. Possibilities that have been put forward include better diagnostic criteria as well as increased awareness but also environmental influences [7]. The etiology of ASD is multifactorial and complex with the interplay of genes in their correlation with the environment. Genetic studies, through twin studies, have shown a significant genetic component involving heritability estimates within a range of 64–91% [8]. The genetic landscape of ASD is, however, highly heterogeneous with involvement of rare as well as common variants scattered across multiple genes.

Pre- and Postnatal environmental factors have been implicated in the risk for ASD. They include advanced parental age, maternal infections in the course of pregnancy, certain medications, or exposure to toxins [9]. In no case, however, has a single environmental factor been demonstrated as a cause specifically for ASD. The condition probably originates from an interaction between genetic predispositions and influences from environment.

9.5 Challenges and Associated Conditions

Although restricted, stereotyped behaviors and impairments in social communication are core features of autism spectrum disorder (ASD), many individuals with ASD also experience a variety of other challenges and comorbid conditions. An intellectual handicap is present in about 31% of people with ASD, indicating a substantial overlap in these developmental problems [4]. Language difficulties are also prevalent; some people either stay barely verbal or experience major delays in language development [10]. Another common co-occurring illness is attention-deficit/hyperactivity disorder (ADHD), whose prevalence among people with ASD varies greatly from 30% to 80% [11]. Another issue is sleeping disturbances, since up to 80% of kids with ASD have trouble falling asleep [12]. Moreover, digestive problems are frequently reported, although research is still being done to determine their exact prevalence and underlying reasons [13]. Finally, it is anticipated that a large portion of people with ASD have a high prevalence of epilepsy—roughly 20%—compared to the general population [14]. A comprehensive approach to therapy and support is required since these other diseases complicate the clinical picture and management of ASD.

9.6 Developmental Trajectory and Outcomes

The developmental trajectory of autism spectrum disorder (ASD) is highly individualized, with outcomes ranging from the need for substantial lifelong support to achieving high levels of independence and success in various fields. Several factors significantly influence these outcomes. Early identification and intervention are crucial; research indicates that diagnosing and addressing ASD early can lead to markedly improved outcomes over the long term [15]. Developmental trajectories are naturally influenced by the intensity of ASD symptoms; people with milder symptoms typically have better outcomes; however, these can vary greatly [16]. Moreover, different secondary problems, such as mental conditions or intellectual incapacities, can exacerbate long-term outcomes. These factors suggest that treatment should be tailored and diagnosed early to address ASD.

9.7 Overview of Play Therapy: Purpose and Importance in ASD

Play therapy represents a therapeutic approach that employs play as the principal medium for communication and expression in a structured setting. This approach is grounded in the understanding that play serves as the inherent language of children, through which they articulate their thoughts, emotions, and experiences [17]. Within the framework of autism spectrum disorder (ASD), the significance of play therapy is amplified, given the distinct challenges and developmental requirements faced by children on the autism spectrum.

The use of the play therapy in ASD is holistic and well-grounded in multiple elements of development and functioning.

- The play therapy is expected to enhance ASD children's social communication and interaction skills, which are core deficiencies. Controlled and free play activities can help ASD children practice and develop essential skills such as turn-taking, joint attention, and social reciprocity in a supportive and engaging environment [18].
- Play therapy is an opportunity for children who are diagnosed with autism spectrum disorder (ASD) to explore and increase their interests and behaviors while utilizing a flexible and adaptive approach. This factor is critical in modifying the restricted and repetitive behaviors that are commonly characteristic of ASD. By introducing new play materials and situations, therapists enable children's interests and foster more adaptable cognitive and behavioral adjustments in them [19].
- It can be used to support the development of emotional regulation and expression in children with ASD. The individuals with ASD battle detecting and communicating their emotions as well as trying to understand the emotions of others. Playing with children can help them realize and express a mixture of emotions in a risk-free, non-threatening environment while gradually increasing emotional awareness and regulation skills [20].
- The play therapy can increase the cognitive, attention, and language development of children with ASD. Playing offers the opportunity for symbolic thinking, problem-solving, and imagination; all of which can be challenging for individuals with ASD. Moreover, the participatory nature of play therapy will help support the child's development of language through both verbal and nonverbal communication [21].

The other important benefit that this form of play therapy can offer ASD children is that it takes advantage of what should be the natural interest and enthusiasm of the child. Incorporation of activities and toys they prefer in therapeutic play has been known to improve engagement and motivation, which are important aspects of any intervention's success [22]. One of the reasons is that play therapy provides a naturalistic setting for the learning and building of skills, which may improve the transfer of skills to real-world situations. This aspect is very important for children with

ASD since they face immense challenges when their acquired skills at structured settings are transferred to everyday environment [5]. The flexibility in application of play therapy enables therapists to shape interventions according to the specific needs and abilities of every child with ASD. This would be adequately situated considering the heterogeneity in ASD presentations, as each person possesses significant strength and challenges [23]. Furthermore, play therapy may also connect the child with ASD to family, sibling, and peers. Incorporating families and peers in play-based interventions, therapists may hence enable positive relationships and foster social learning and interaction within a protective environment.

In the direct benefits bestowed on children with ASD, play therapy also avails insights into the inner world of the child. Thus, for therapists and carers, it is of immense help in better understanding their unique perspective and experiences. It could be a basis for formulating more appropriate support strategies which also would represent greater empathy besides being well worked out within the therapy sessions [24]. Still, it is crucial to highlight that play therapy is very promising for children with ASD, though it should never be used as a solitary intervention. It should, however, be part of a general treatment plan that can include some other evidence-based interventions, such as cognitive-behavioral therapy, speech and language therapy, and occupational therapy [25]. Additional strategies employed for autism spectrum disorder (ASD) include technology-enhanced play interventions, which utilize digital tools to facilitate social interaction and promote skill development, as well as nature-based play therapies, which capitalize on the advantages of outdoor environments to support sensory regulation and enhance social engagement [26].

9.8 Stages of Childhood and Impact of Play Therapy

Play therapy has different benefits at various stages of childhood and seriously affects child development pathways. The following section outlines its varying effects:

9.8.1 Intrauterine Fetal Play

In the intrauterine phase, play therapy encompasses sensory stimulation techniques, including the gentle caressing of the mother's abdomen, verbal communication, and the playing of soothing music aimed at the fetus. This method of early interaction promotes sensory growth and fosters the formation of initial emotional connections. It develops the tactile and auditory systems of the fetus, which could be a long-term modifier of early cognitive functions, such as the ability to recognize language and process emotional information [27]. Touching the fetus in these ways could strengthen the emotional attachment and provide a foundation for sensory pathways to continue developing after birth.

9.8.2 Neonatal Play

In the neonatal stage, play therapy typically involves sensory and social interaction. Such activities include tactile stimulation with soft toys, auditory involvement through soft speaking or singing, and visual stimulation with simple, high-contrast pictures or books [28]. These interactions help establish a safe attachment between the infant and their caregivers, which is essential in developing feelings of emotional security and trust. In addition, these early playful interactions foster the development of both fine and gross motor skills and create a basis for early speech and language skills through repetition of sounds.

9.8.3 Early Childhood Play

Entering into early childhood, play therapy remains the primary facilitator of more challenging achievements. It helps the child better develop social skills, emotional regulation, and cognitive powers. Structure in play sessions allow children to experiment on the expression of emotions and conflict resolution besides fostering empathy in cooperative peer activities [29]. Cognitive skills are further entrenched through problem-solving tasks involved in play, and it improves creativity for enhancing readiness for academics. Physical play will engender health, strength, and coordination; cultural play develops children's understanding of, and ability to cope with, their new roles in society. During this stage of play therapy, the communication skills learned, for effective learning and social interaction, are consolidated [29].

9.8.4 Late Childhood Play

Play therapy at late childhood, from about 9 to 12 years, takes a different approach since the child's needs have become quite complex during adolescence. The focus at this stage often shifts toward enhancing problem-solving skills and emotional intelligence. Play activities such as strategic board games or complex building tasks help in developing advanced cognitive skills, including critical thinking and logical reasoning [30]. Social development is also a key focus, with group play scenarios providing opportunities for children to practice negotiation, leadership, and empathy. This helps them navigate more complex social structures and relationships. Additionally, play therapy during late childhood encourages children to explore and express complex emotions through activities like drama and role-play, aiding in emotional maturity [31]. Physical play remains important, with sports and activities helping to refine motor skills and promote physical health, preparing them for the physical changes of adolescence.

9.8.5 Adolescent Play

As children transition into adolescence, play therapy continues to play a crucial role in addressing their evolving challenges and needs before they become adults [4]. In play therapy, the strategy for the approach used with adolescents is to place more emphasis on identity exploration and social maturity. Role-playing games and stories can help teenagers explore various aspects of their identity, thus finalizing interest, belief, and value in a safe expressive environment. It also helps with the emotional growth of adolescents who learn to cope with their complicated feelings and build resilience to social forces that press upon them to oppress. Social aspects of play therapy are improvements of communication abilities, understanding social cues, and development of deeper relationships that are important in the fact that in adulthood they prepare for more independent social environments. In general, play therapy during adolescence lays the groundwork for emotional stability and social readiness, essential to navigation through adult life's challenges [32].

All these phases indicate that play therapy can be specifically designed to cater to the development of needs of the children in ways that result in growth and adaptation through planned and free-play activities. Such versatility makes play therapy a powerful tool to be used in a wide range of developmental accomplishments. A detailed summary of the stages of play therapy, their objectives, activities, and progression levels, is provided in Table 9.4

9.9 Historical Development of Play Therapies for Autism

The history of play therapies designed for people with autism also follows closely in step with the growing understanding of autism spectrum disorders and the broader field of play therapy. In order to fully appreciate the state of play therapy for autism today, it becomes necessary to trace its historical origins and the landmark events that have shaped its use. The idea of using play as an entrance route for therapy of children has its roots in the early twentieth century, and innovators like Anna Freud and Melanie Klein recognized the role of play in understanding and treating children's psychological problems [33]. However, the specific application of play therapy to individuals diagnosed with autism emerged later, coinciding with progress in understanding autism itself.

In the early years post such descriptions, autism was often understood through a psychoanalytic lens: Some theorists attributed it to cold, unresponsive parenting—a notion since thoroughly debunked [34]. This perspective led to early approaches to therapy for autism, including play therapy being framed within it. By the 1960s and 1970s the direction for developing treatment approaches for autism turned toward behaviorist techniques, especially where applied behavior analysis established itself as one of the major interventions [35]. Even so, though these approaches were often effective, the emotional and relational aspects of developmental factors frequently were left without support. It was against this background that therapy through play for autism started taking shape as a unique approach. In the 1970s and 1980s, the researchers and clinicians recognized the scope of play-based interventions to meet the needs of children with unique conditions of autism. Stanley Greenspan is one of the leading figures in this field who was first to create the Developmental, Individual-difference, Relationship-based model, in familiar words, known as Floortime [36].

The DIR/Floortime approach speaks to leading the child in play and playful interactions in order to support a child's ability to develop emotionally and socially. This is quite different than more structured behavioral approaches but it does emphasize relational and relationship-based intervention approaches for autism. Around this time, other play-based approaches for autism were being developed as well. Pamela Wolfberg developed the Integrated Play Groups (IPG) model in the 1980s to support children with autism in engaged peer play and social interaction [37]. This model proved the need for the child with ASD to maintain relationships with their peers and a social context for play in supporting growth.

The attention to the evidence-based practices in the treatment of autism increased in the 1990s, which included play-based approaches. Researchers have become more rigorous in their studies pertaining to the effectiveness of play interventions with children diagnosed with autism. Some researchers have studied an example of a joint attention intervention designed using play-based strategies in order to enhance social communication skills in young children with autism [37]. Compressed understanding of autism as a spectrum condition in the late 1990s and early 2000s led to a boom in the diversity of play therapies intended for autism that were tailored to meet certain individual requirements. Acknowledgment of the variability in manifestations contributed to the development of more elastic and personalized play-centered therapeutic approaches.

Lately, significant interest has been growing in the application of technology in play therapy for autism. Digital tools, virtual reality, and robotics have opened up vast vistas of potential interactions of children with ASD in therapeutic play [38]. These novel technologies are likely to add value to existing means of play therapies, opening new avenues for skill-building and socialization skills. Another recent trend in play therapy for autism is the increased focus on sensory integration and regulation. Recognizing the sensory processing differences often present in individuals with ASD; many play therapy approaches now incorporate sensory-based activities and strategies to support regulation and engagement [39].

There is a history to play therapies for autism, suggesting how our understanding of autism spectrum disorders has changed. From the early psychoanalytic methods, the treatment has moved through behavioral treatments and later on to more combined, developmental, and relationship-based approaches. In other words, the practice has been constantly evolving with new research and clinical knowledge. Presently, play therapy for autism has turned out to be a diversified and active field, which combines different techniques into one amalgamation of multiple ideas and treatment approaches. As much as there is still a lot to explore in the best ways of applying play therapy among people with ASD, the history of these techniques has already established a core foundation for further research and new clinical ideas.

As the domain of play therapy for autism progresses, it is characterized by continual evolution, with persistent research aimed at improving current methodologies and formulating innovative techniques. The amalgamation of knowledge derived from neuroscience, developmental psychology, and other pertinent disciplines holds the potential to significantly deepen our comprehension of the ways in which play-oriented interventions can foster the growth and welfare of those diagnosed with autism spectrum disorders.

9.10 Types of Play Therapy Used in Autism

Play therapy is one of the essential intervention approaches for children diagnosed with autism spectrum disorder, providing a flexible, engaging method to overcome the core challenges of the condition. This section will explore various types of play therapy techniques used in autism intervention, including directive and nondirective approaches, music therapy combined with play, parent-mediated and full-body interaction interventions, and psychodynamic play therapy. A summary of these play types and their impacts is presented in Table 9.1.

Table 9.1 Types of play therapy and their impact

Type of play	Description	Impact
Object play	Active manipulation of objects begins around 5 months and evolves into complex constructive play by age 2	Enhances visual-motor integration, cognitive development, social interaction, and motor skills
Pretend play	Starts by 12 months, involving imaginary scenarios and role-playing with objects symbolizing real items	Boosts creativity, social and language skills, emotional regulation, and executive functions
Physical play	Involves physical activities like running and climbing, peaking in early childhood and tapering off in adolescence	Promotes physical health, coordination, cognitive development, and sensory integration
Media play	Engagement with digital media and technology-based games, increasing in hours from early childhood	Affects cognitive, social, and physical development variably; enhances learning and visuospatial skills in older children

9.11 Directive vs. Nondirective Play Therapy

Play therapy approaches can be broadly categorized into directive and nondirective techniques, each with its own set of principles and applications in the context of autism intervention.

9.11.1 Nondirective Play Therapy

Nondirective play therapy, also known as child-centered play therapy, is rooted in the work of Virginia Axline and her adaptation of Carl Rogers' person-centered approach to working with children [40]. This approach is based on the premise that children have an innate drive toward growth and self-actualization and that, given the right conditions, they can resolve their own emotional and behavioral difficulties through play.

In nondirective play therapy, the therapist creates a safe and accepting environment where the child is free to express themselves through play without guidance or interpretation from the therapist. The therapist's role is to provide empathic understanding, unconditional positive regard, and reflective responses to the child's play and emotions [41].

For children with ASD, nondirective play therapy can offer several benefits:

1. The first benefit for children with ASD is the experience of the world where they have limited control over their environment and daily routines. The children can take a lead and make choices in the nondirective play, potentially fostering a sense of autonomy and self-efficacy [42].
2. Many children with ASD find it difficult to identify and express emotions. The nonjudgmental and accepting nature of nondirective play therapy can create a safe space for these children to explore and express their feelings through play [43].
3. The nondirective play therapy can improve the flexibility in thinking and behavior. The open-ended nature of nondirective play can encourage children with ASD to explore new ways of interacting with objects and ideas, potentially promoting cognitive flexibility [44].
4. The nondirective play therapy helps in relationship building with the child and therapist, which can provide a model for positive social interactions, which is particularly valuable for children with ASD who often struggle with social relationships [45].

The effectiveness of nondirective play therapy for children with ASD may be constrained by the inherent challenges of the condition. For instance, children with ASD often struggle with symbolic play and initiating social interactions, both of which are crucial aspects of nondirective play therapy [37].

9.11.2 Directive Play Therapy

Directive play therapy involves a more structured approach where the therapist takes an active role in guiding the play session and selecting activities. This approach is often based on specific therapeutic goals and may incorporate elements of other intervention strategies, such as behavioral or cognitive-behavioral techniques [46].

For children with ASD, directive play therapy can offer several advantages:

1. Structured learning opportunities: Therapists can create play activities designed to address specific skills commonly needed in children with ASD, such as joint attention, turn-taking, and social communication [47].
2. Clear expectations: Children with ASD often benefit from predictable structures. Directive play therapy can offer clear guidelines and expectations for the session, which may help reduce anxiety and increase participation [48].
3. Skill generalization: By integrating naturalistic teaching strategies into play, directive play therapy can help children generalize learned skills to everyday situations [49].
4. Addressing specific behaviors: This therapy approach can be customized to target specific behavioral issues associated with ASD, such as repetitive actions or restricted interests [50].

9.12 Examples of Directive Play Therapy Approaches

One notable approach is Integrated Play Groups (IPG), developed by Pamela Wolfberg. IPG organize structured play sessions that include both children with ASD and their typically developing peers, with adult guidance to support social interactions and play skills [37]. Another directive method is Theraplay, which focuses on improving attachment, self-esteem, and trust through playful, physical interactions between the child and a therapist or parent [51]. Although Theraplay was not originally designed for children with ASD, studies suggest that its techniques can be beneficial, especially in enhancing social engagement and reducing problematic behaviors [52].

It is important to recognize that the line between directive and nondirective play therapy is often blurred, with many practitioners adopting a flexible approach that blends both styles. For instance, the Developmental, Individual-difference, Relationship-based (DIR)/Floortime model, created by Stanley Greenspan, incorporates both child-led and therapist-guided strategies to foster social-emotional growth in children with ASD [36]. Another study compared child-centered play therapy with behavioral-based play therapy in children with autism. The findings indicated that both methods improved social skills and reduced problem behaviors, though the directive (behavioral-based) approach demonstrated more significant gains in social skills. A summary of these therapy types, their descriptions, effectiveness, and examples are provided in Table 9.2.

Table 9.2 Types of play therapy, description, and examples of activities

Therapy type	Approach description	Effectiveness	Examples/specific outcomes	Examples of games/activities
Child-centered play therapy	Relationship-based interaction where therapists engage with children in their own world	Reduces core ASD symptoms like attention problems and aggression; improves joint attention and social skills	It reduces the externalizing problems and aggression	Sand play, dollhouses, and art materials allowing children to express emotions and scenarios freely
Child-led play	Nondirective, allowing children to choose activities, enhancing autonomy and interaction	Effective in severely affected cases; long-term benefits in social communication skills	It reports improvements in social communication skills after using pivotal response treatment	Activities chosen by the child, such as building blocks, puzzles, or drawing, facilitating child's choice and control over the play environment
Floortime-play therapy	Therapists follow the child's lead in play, gradually introducing new elements to foster growth	Improves attention, engagement, and social interactions; requires extensive parental involvement	Children shows the significant progress in functional development with regular floortime sessions	Role-playing with toys, imaginative scenarios, and interactive games that encourage back-and-forth conversation and interaction
Clown therapy	Uses humor and play to alleviate stress and anxiety, adapting to hospital settings	Shows potential in enhancing communication and reducing ASD-related behaviors in clinical settings	It is observed the reduction in stereotypic movement frequency during medical clowning sessions	Playful interactions with magic tricks, juggling, or funny routines that engage children and reduce anxiety and stress
3i play therapy	Intensive, individualized, and interactive approach focusing on daily extensive interaction at home	Significant behavioral and developmental improvements; enhances daily interactions and social environment adaptability	It reports a decrease in autism severity after 2 years of intensive daily play sessions	Use of everyday objects in creative ways, like turning a box into a spaceship, which encourages imaginative play and social interaction

(continued)

Table 9.2 (continued)

Therapy type	Approach description	Effectiveness	Examples/specific outcomes	Examples of games/activities
Integrated play groups	Groups neurotypical peers with children with ASD to improve social skills through facilitated interactions	Enhances symbolic and social play; increases peer interaction and social inclusion	The symbolic play and social interaction in children participation can be improved	Games that require turn-taking and cooperation, such as board games or group sports, which foster social interactions and shared goals
JASPER	Focuses on improving joint attention and play skills through structured interventions	Enhances joint engagement, language skills, and play diversity; effective in diverse settings including online delivery	The children showed greater diversity in play and engagement in classroom settings after JASPER interventions	Structured play activities like "I Spy" or treasure hunts that encourage children to use joint attention and shared engagement
Synergetic play therapy	Incorporates all elements of a child's life to create a holistic therapeutic environment	Induces emotionally tolerant behaviors; promotes overall child development and emotional resilience	It helps in the improvement in emotionally tolerant behaviors in children from the first to third sessions of synergetic play therapy	Activities that reflect real-life challenges, such as obstacle courses or role-play scenarios, which help children manage emotions and interactions
Equine-partnered play therapy	Combines equine-assisted therapy with play to exploit the therapeutic potential of animal interactions	Improves sensory seeking, social motivation, and reduces distractibility; enhances motor control and self-care	It improves the social functioning and reduced hyperactivity in children after weekly horse-riding sessions	Horse-riding and grooming activities that promote bonding with animals and develop responsibility, empathy, and nonverbal communication skills
Art-play therapy	Uses art as a medium of expression to address psychological conflicts and foster communication	Positively impacts speech, social relations, and behaviors; enhances sensory experiences and social behavior	The art therapy sessions led to enhanced speech and social interactions among children and adolescents with ASD	Painting, clay modeling, or collage making that allows children to express their feelings visually and improve fine motor skills

(continued)

Table 9.2 (continued)

Therapy type	Approach description	Effectiveness	Examples/specific outcomes	Examples of games/activities
Tele-play therapy	Virtual therapy aimed at providing therapeutic services remotely, especially during restrictive conditions	Facilitates continued therapeutic engagement; requires careful consideration of privacy and technological challenges	Implemented widely during the COVID-19 pandemic, allowing continued therapeutic support despite lockdowns, reducing isolation and enabling ongoing developmental support	Digital puzzles, interactive story-telling, and online drawing sessions that engage children remotely while maintaining therapeutic goals

9.13 Music Therapy as an Adjunct to Play Therapy

Music therapy has gained recognition as a valuable intervention for individuals with ASD, often incorporating elements of play to enhance engagement and target specific developmental goals. The combination of music and play can create a multisensory experience that capitalizes on the often-observed affinity for music in individuals with ASD [53].

Music therapy interventions for children with ASD typically involve interactive music-making, often using instruments, singing, and movement. These activities can be structured to address various areas of development, including:

1. Social interaction and communication: Music-making naturally encourages turn-taking, joint attention, and nonverbal communication, which are often challenging for children with ASD [25]
2. Emotional regulation: The rhythmic and structured nature of music can help children with ASD regulate their emotions and behavior [54].
3. Motor skills: Playing instruments and engaging in music-based movement activities can support the development of fine and gross motor skills [55].
4. Cognitive skills: Cognitive skills such as attention, memory, and sequencing can be improved with music activities [56].
5. Language development: Singing and rhythmic speech activities can support language acquisition and expressive communication [57].

Several specific music therapy approaches have been developed or adapted for use with children with ASD:

9.13.1 Improvisational Music Therapy

This method involves the therapist and child making music together spontaneously, using instruments or their voices. The therapist tunes into the child's musical expressions and responds in ways that encourage interaction and communication. A study has looked at the effects of improvisational music therapy on social affect in children with ASD. Although they did not find significant differences between the music therapy and standard care groups on the main outcome measure, they observed that music therapy improved parent-reported social responsiveness in a subset of children [58].

9.13.2 Neurologic Music Therapy (NMT)

NMT is a research-based system of standardized clinical techniques that use music to target nonmusical goals in sensorimotor, speech/language, and cognitive domains. LaGasse (2014) conducted a study using NMT techniques to improve social skills in children with ASD [55]. The results showed improvements in eye gaze and joint attention behaviors following the intervention.

9.14 Auditory-Motor Mapping Training (AMMT)

AMMT is a specialized technique that integrates singing and motor activities to encourage speech production in nonverbal children with ASD. Researchers discovered that AMMT resulted in notable improvements in speech output among previously non-verbal children with ASD [59].

9.15 The DRUMBEAT Program

This group-based intervention combines hand drumming with social learning activities. While not specifically developed for ASD, Fancourt et al. (2016) found that participation in DRUMBEAT was associated with improvements in social skills and reductions in anxiety symptoms in a group of adolescents with ASD [60].

9.16 Music-Based Play Interventions

Some researchers have developed interventions that explicitly combine music and play elements. For example, musical interaction therapy is a technique, which involves playful, music-supported interactions between a child with ASD and their caregiver. They reported improvements in eye contact, child-initiated interactions, and joint attention following the intervention [61].

Studies have shown that integrated music strategies into classroom play routines to enhance peer interactions and meaningful play among young children with ASD [62]. Their findings indicated that these music interventions led to increased social interactions and more engagement in play activities. While the overall body of research supports the utility of music therapy for children with ASD, it is crucial to acknowledge that the evidence quality is mixed.

9.17 Parent-Mediated and Full-Body Interaction Interventions

9.17.1 Parent-Mediated Interventions

Parent-mediated interventions have gained increasing attention in the field of autism treatment, recognizing the crucial role that parents play in their child's development and the potential for intervention effects to generalize to daily life when parents are actively involved [63].

In the realm of play therapy, parent-mediated interventions involve educating parents on how to apply play-based techniques to aid their child's development. These methods frequently incorporate concepts from other proven interventions, such as naturalistic developmental behavioral interventions (NDBIs), modifying them for use during parent-child play sessions [64]. A detailed comparison of parent-mediated play interventions, including their descriptions and outcomes, is presented in Table 9.3.

Table 9.3 Parent-mediated play interventions for children with ASD

Intervention	Description	Outcome improvements
JASPER (Joint Attention, Symbolic Play, Engagement, and Regulation) [65]	A play-based intervention targeting deficits in joint attention, symbolic play, and social interaction. Parents guide the child's play using specific strategies	Enhanced joint engagement, joint attention, and play skills
DIR/Floortime [66]	Parents engage in child-led play that supports social-emotional development and relationship building	Improvements in social interaction, communication, and adaptive behavior
Pivotal response treatment (PRT) [67]	Focuses on motivation and responsiveness to multiple cues, with parent-implemented versions available	Enhancements in language, play skills, and social behavior
Early Start Denver Model (ESDM) [68]	An early intervention program that instructs parents in naturalistic, play-centered methods	Enhanced social communication skills and adaptive behaviors

9.18 Benefits of Parent-Mediated Interventions

- Increased intervention dosage: Allows for more frequent and integrated intervention sessions via daily activities.
- Enhanced skill generalization: Skills are more likely to be applied in natural, everyday settings.
- Improved parent-child relationships: Can lead to stronger bonds and reduced parental stress.
- Cost-effectiveness: More affordable compared to interventions requiring professional delivery.

9.19 Challenges

- Treatment fidelity: Ensuring parents correctly implement therapeutic techniques
- Parental stress and burnout: Managing the emotional and physical demands on parents
- Adaptation to family routines: Integrating interventions seamlessly into daily life
- Cultural adaptations: Customizing interventions to suit various cultural contexts

9.20 Full-Body Interaction Interventions

Full-body interaction interventions for children with ASD focus on engaging the entire body in play-based activities to promote various developmental skills. These approaches recognize the importance of sensorimotor experiences in learning and development, as well as the potential benefits of physical activity for individuals with ASD [69].

Some examples of full-body interaction interventions include:

1. *Movement-based learning*

This approach incorporates whole-body movements into learning activities. For example, Tiffany Field and her colleagues have investigated the benefits of movement therapy for children with ASD, discovering that movement-based interventions lead to better social connections and increased attentiveness [70].

2. *Exergaming*

Exergames are video games that require physical movement to play. Some researchers have investigated the use of exergames to encourage physical activity and foster social interaction in children with ASD. The researchers found that children with ASD showed increased social engagement when playing exergames with peers compared to sedentary video games [71].

3. *Dance/movement therapy*

This creative arts therapy uses movement to promote emotional, social, cognitive, and physical integration. While research specifically on dance/movement therapy for ASD is limited, some studies have reported positive outcomes, such as improved social skills and reduced stereotypical behaviors [72, 73].

4. *Yoga and mindfulness-based interventions*

These approaches incorporate full-body movements with breath awareness and mindfulness techniques. Few studies have reported development in social skills, attention, and behavioral regulation in children with ASD following yoga interventions [74].

5. *Play-based physical activity interventions*

These methods integrate playful elements into physical activity programs tailored for children with ASD. It is seen that a 10-week swimming exercise program results in enhanced aquatic skills and social behaviors in children with ASD [75].

The potential benefits of full-body interaction interventions for children with ASD include:

- Improved motor skills and physical fitness [69]
- Enhanced sensory processing and integration [46]
- Increased social interaction and communication opportunities [66]
- Reduced stereotypical behaviors and improved attention [25]

9.21 Psychodynamic Play Therapy

Psychodynamic play therapy is an approach that draws on psychoanalytic principles to understand and treat emotional and behavioral difficulties in children. This approach views play as a means of communication and a window into the child's inner world, allowing for the expression and working through of unconscious conflicts and emotions [76].

While psychodynamic play therapy was not originally developed for children with ASD, some clinicians and researchers have explored its potential applications for this population. The use of psychodynamic play therapy with children with ASD is based on the premise that these children, like all children, have inner emotional lives and unconscious processes that influence their behavior and development [77].

Key principles of psychodynamic play therapy as applied to children with ASD include:

1. Emphasis on the therapeutic relationship: The therapist aims to develop a secure, containing relationship with the child, which can serve as a basis for emotional growth and social learning [78].
2. Attention to symbolic communication: The therapist looks for meaning in the child's play, behaviors, and interactions, even when these may appear repetitive or nonsymbolic [77].
3. Focus on emotional development: The therapy aims to support the child's emotional awareness, expression, and regulation [79].
4. Consideration of defensive mechanisms: The therapist considers how certain behaviors or play themes might serve as defenses against anxiety or overwhelming emotions [80].
5. Exploration of internal object relations: The therapy examines how the child's internal representations of self and others influence their interactions and behaviors [77].

Some psychodynamic practitioners have developed specific approaches for working with children with ASD:

1. *Anne Alvarez's "Live Company" approach*

Alvarez (1992) emphasizes the importance of the therapist's active, emotionally engaged presence in working with children with ASD. She describes techniques for "reclaiming" children from states of withdrawal and helping them develop a sense of relatedness [77].

2. *Frances Tustin's work on autistic shapes*

Tustin (1991) explored the sensory experiences of children with ASD and how these might relate to early developmental processes. Her research influences therapeutic strategies that focus on the sensory components of a child's play and behavior [80].

3. *Didier Houzel's "Envelope Theory"*

Houzel (2004) proposed the concept of a "psychic envelope" that helps contain and organize sensory experiences and emotions. His research has shaped therapeutic methods designed to foster the development of emotional regulation in children with ASD [81].

It is important to note that psychodynamic play therapy for children with ASD often incorporates elements from other approaches and may be adapted to address the specific needs and challenges associated with autism. For example:

1. Increased structure: While traditional psychodynamic play therapy is often nondirective, work with children with ASD may involve more structure and active engagement from the therapist [78].

2. Focus on concrete experiences: The therapist may use more concrete language and experiences to support the child's understanding and engagement [77].
3. Attention to sensory needs: The therapy may incorporate sensory elements to support regulation and engagement [79].
4. Integration with other approaches: Psychodynamic play therapy may be used in combination with other evidence-based interventions for ASD, such as behavioral or developmental approaches [78].

Potential benefits of psychodynamic play therapy for children with ASD include:

1. Supporting emotional development: The therapy can help children with ASD become more aware of and expressive with their emotions [79].
2. Enhancing relational capacities: Through the therapeutic relationship, children may develop greater capacity for social connection and understanding [77].
3. Addressing underlying anxieties: The therapy can help identify and work through anxieties that may underlie certain behaviors or difficulties [80].
4. Promoting symbolic thinking: Through play and interpretation, the therapy may support the development of symbolic thought and pretend play [15].
5. Individualizing treatment: Psychodynamic approaches emphasize understanding each child's unique internal world, which can lead to highly individualized interventions [78].

However, it is important to acknowledge the limitations and challenges of using psychodynamic play therapy with children with ASD:

1. Limited empirical evidence: Compared to other intervention approaches for ASD, there is less rigorous research supporting the effectiveness of psychodynamic play therapy [82].
2. Potential mismatch with ASD characteristics: Some core features of ASD, such as difficulties with symbolic thinking or theory of mind, may present challenges for traditional psychodynamic approaches [83].
3. Time-intensive nature: Psychodynamic therapy often involves long-term treatment, which may not be feasible or cost-effective for all families [84].
4. Therapist expertise: Effective implementation requires specialized training in both psychodynamic approaches and ASD, which may limit availability [78].
5. In summary, while psychodynamic play therapy offers a unique method for enhancing the emotional and social development of children with ASD, additional research is necessary to validate its effectiveness and identify the children who will benefit the most. As with any ASD intervention, it is essential to consider the unique needs and traits of each child when choosing the most appropriate therapeutic approach (Table 9.4).

Table 9.4 Stages of play therapy, description, objectives, and progression

Stage	Description	Objectives	Key activities	Duration	Aim	Level for progression
Stage 1	Initial engagement	Establish rapport and a safe therapeutic environment	Introduce simple, nonthreatening play activities. Allow the child to explore the play area and choose toys. Observe and reflect the child's actions and emotions	2–4 weeks	To make the child feel safe and understood	Child regularly attends sessions and begins to initiate play
Stage 2	Deepening engagement and building trust	Deepen the therapeutic relationship and trust between therapist and child	Engage in more interactive play such as turn-taking games. Begin to introduce activities that require joint attention. Continue to mirror and validate the child's feelings	1–2 months	To foster trust and begin interactive play	Child starts interacting with the therapist and shows comfort in the play setting
Stage 3	Addressing core ASD symptoms	Use play to address specific symptoms of ASD such as social communication challenges, rigid behaviors, and sensory integration issues	Incorporate structured play to teach social skills like sharing and cooperation. Use sensory play activities to aid sensory integration. Implement role-play and pretend play to enhance flexibility and imagination	3–6 months	To modify behaviors and enhance skills	Child demonstrates new skills in play and begins to apply them in interactions
Stage 4	Generalization and integration	Help the child apply learned skills in broader contexts	Introduce play scenarios that mimic real-life interactions. Encourage play with peers in a supervised setting to practice social skills. Plan activities that involve family members to reinforce skills at home	6–8 months	To stabilize and generalize skills	Child uses learned skills in various settings and maintains improvements over time

9.22 Integrating Different Play Therapy Approaches

While we have examined various forms of play therapy individually, it is essential to recognize that, in clinical practice, many therapists use an integrative approach. This approach blends elements from different play therapy methods to better suit the unique needs of each child with ASD. This integrative strategy acknowledges the diverse nature of ASD and the reality that no single intervention can address all the challenges associated with the condition [49].

References

1. Lai M-C, Lombardo MV, Baron-Cohen S. Search strategy and selection criteria. Lancet. 2014;383:896–910.
2. Warren Z, McPheeters ML, Sathe N, Foss-Feig JH, Glasser A, Veenstra-VanderWeele J. A systematic review of early intensive intervention for autism spectrum disorders. Pediatrics. 2011;127(5):e1303–e11.
3. Tiede G, Walton KM. Meta-analysis of naturalistic developmental behavioral interventions for young children with autism spectrum disorder. Autism. 2019;23(8):2080–95.
4. van Engeland H, Buitelaar JK. Autism spectrum disorders. In: Rutter's child and adolescent psychiatry. 2008. p. 759–81.
5. Young RL, Rodi ML. Redefining autism spectrum disorder using DSM-5: the implications of the proposed DSM-5 criteria for autism spectrum disorders. J Autism Dev Disord. 2014;44:758–65.
6. Knott F, Dunlop A-W, Mackay T. Living with ASD: how do children and their parents assess their difficulties with social interaction and understanding? Autism. 2006;10(6):609–17.
7. Bougeard C, Picarel-Blanchot F, Schmid R, Campbell R, Buitelaar J. Prevalence of autism spectrum disorder and co-morbidities in children and adolescents: a systematic literature review. Front Psychiatry. 2021;12:744709.
8. Tick B, Bolton P, Happé F, Rutter M, Rijsdijk F. Heritability of autism spectrum disorders: a meta-analysis of twin studies. J Child Psychol Psychiatry. 2016;57(5):585–95.
9. Ornoy A, Weinstein-Fudim L, Ergaz Z. Prenatal factors associated with autism spectrum disorder (ASD). Reprod Toxicol. 2015;56:155–69.
10. Vogindroukas I, Stankova M, Chelas E-N, Proedrou A. Language and speech characteristics in autism. In: Neuropsychiatric disease and treatment. 2022. p. 2367–77.
11. Carta A, Fucà E, Guerrera S, Napoli E, Valeri G, Vicari S. Characterization of clinical manifestations in the co-occurring phenotype of attention deficit/hyperactivity disorder and autism spectrum disorder. Front Psychol. 2020;11:861.
12. Schreck KA, Mulick JA. Parental report of sleep problems in children with autism. J Autism Dev Disord. 2000;30:127–35.
13. Lefter R, Ciobica A, Timofte D, Stanciu C, Trifan A. A descriptive review on the prevalence of gastrointestinal disturbances and their multiple associations in autism spectrum disorder. Medicina. 2019;56(1):11.
14. Canitano R. Epilepsy in autism spectrum disorders. Eur Child Adolesc Psychiatry. 2007;16:61–6.
15. Camarata S. Early identification and early intervention in autism spectrum disorders: accurate and effective? Int J Speech-Lang Pathol. 2014;16(1):1–10.
16. Baghdadli A, Assouline B, Sonié S, Pernon E, Darrou C, Michelon C, et al. Developmental trajectories of adaptive behaviors from early childhood to adolescence in a cohort of 152 children with autism spectrum disorders. J Autism Dev Disord. 2012;42:1314–25.
17. Knell SM. Cognitive-behavioral play therapy. Handbook of play therapy. 2015. p. 119–33.

18. Hart Barnett J. Three evidence-based strategies that support social skills and play among young children with autism spectrum disorders. Early Childhood Educ J. 2018;46(6):665–72.
19. Kendall PC. Guiding theory for therapy with children and adolescents. In: Child and adolescent therapy: cognitive-behavioral procedures. 1991. p. 2.
20. Saleem S. Social skills among aggressive and non-aggressive children. Quaid I Azam University; 2006.
21. Preissler MA. Play and autism: facilitating symbolic understanding. In: Play= learning: how play motivates and enhances children's cognitive and socio-emotional growth. 2006. p. 231–50.
22. Holdsworth E, Bowen E, Brown S, Howat D. Client engagement in psychotherapeutic treatment and associations with client characteristics, therapist characteristics, and treatment factors. Clin Psychol Rev. 2014;34(5):428–50.
23. Elbeltagi R, Al-Beltagi M, Saeed NK, Alhawamdeh R. Play therapy in children with autism: its role, implications, and limitations. World J Clin Pediatr. 2023;12(1):1.
24. Tilmont Pittala E, Saint-Georges-Chaumet Y, Favrot C, Tanet A, Cohen D, Saint-Georges C. Clinical outcomes of interactive, intensive and individual (3i) play therapy for children with ASD: a two-year follow-up study. BMC Pediatr. 2018;18:1–13.
25. Will MN, Currans K, Smith J, Weber S, Duncan A, Burton J, et al. Evidenced-based interventions for children with autism spectrum disorder. Curr Probl Pediatr Adolesc Health Care. 2018;48(10):234–49.
26. Koumpouros Y, Kafazis T. Wearables and mobile technologies in Autism Spectrum Disorder interventions: a systematic literature review. Res Autism Spectr Disord. 2019;66:101405.
27. Salo S, Lampi H. Prenatal and infant therapaly. In: Theraplay®-theory, applications and and implementation. Jessica Kingsley Publishers; 2020. p. 83–100.
28. Besio S. Play and disabilities. In: Robot play for all: developing toys and games for disability: Springer; 2022. p. 75–150.
29. Jamero JLF. Social constructivism and play of children with autism for inclusive early childhood. Int J Early Childhood Special Educ. 2019;11(2):154–67.
30. Rutherford MD, Young GS, Hepburn S, Rogers SJ. A longitudinal study of pretend play in autism. J Autism Dev Disord. 2007;37:1024–39.
31. Rao Z, Gibson J. The role of pretend play in supporting young children's emotional development. In: The SAGE handbook of developmental psychology and early childhood education. 2019. p. 63–79.
32. Weisz JR, Hawley KM. Developmental factors in the treatment on adolescents. J Consult Clin Psychol. 2002;70(1):21.
33. Muchacka B. Play as a form of therapy for children (historical and pedagogical aspects of problem). 2013.
34. Waters L. Reinforcing the empty fortress: an examination of recent research into the treatment of autism. Educ Stud. 1990;16(1):3–16.
35. Lovaas OI. Behavioral treatment and normal educational and intellectual functioning in young autistic children. J Consult Clin Psychol. 1987;55(1):3.
36. Greenspan SI, Wieder S. Engaging autism: using the floortime approach to help children relate, communicate, and think. Da Capo Lifelong Books; 2006.
37. Wolfberg P, DeWitt M, Young GS, Nguyen T. Integrated play groups: Promoting symbolic play and social engagement with typical peers in children with ASD across settings. J Autism Dev Disord. 2015;45:830–45.
38. Qazi S, Raza K. Towards a VIREAL platform: virtual reality in cognitive and behavioural training for autistic individuals. In: Advanced computational intelligence techniques for virtual reality in healthcare. 2020. p. 25–47.
39. Case-Smith J, Weaver LL, Fristad MA. A systematic review of sensory processing interventions for children with autism spectrum disorders. Autism. 2015;19(2):133–48.
40. Bratton S, Purswell K, Jayne K. Play therapy: a child-centered approach. In: Counseling and psychotherapy with children and adolescents: theory and practice for school and clinical settings, 5th ed. 2015. p. 91–113.

41. Stulmaker HL, Lertora I, Garza Y. Facilitating empathic perspective taking in beginning child-centered play therapists: the role of supervision. Int J Play Ther. 2015;24(4):177.
42. Salter K, Beamish W, Davies M. The effects of child-centered play therapy (CCPT) on the social and emotional growth of young Australian children with autism. Int J Play Ther. 2016;25(2):78.
43. Hillman H. Child-centered play therapy as an intervention for children with autism: a literature review. Int J Play Ther. 2018;27(4):198.
44. Mastrangelo S. Play and the child with autism spectrum disorder: from possibilities to practice. Int J Play Ther. 2009;18(1):13.
45. Kossyvaki L, Papoudi D. A review of play interventions for children with autism at school. Int J Disabil Dev Educ. 2016;63(1):45–63.
46. Schaefer CE. Foundations of play therapy. Wiley; 2011.
47. Ingersoll B, Schreibman L. Teaching reciprocal imitation skills to young children with autism using a naturalistic behavioral approach: effects on language, pretend play, and joint attention. J Autism Dev Disord. 2006;36:487–505.
48. Ware Balch J, Ray DC. Emotional assets of children with autism spectrum disorder: a single-case therapeutic outcome experiment. J Couns Dev. 2015;93(4):429–39.
49. Stahmer AC, Schreibman L, Cunningham AB. Toward a technology of treatment individualization for young children with autism spectrum disorders. Brain Res. 2011;1380:229–39.
50. Jung S, Sainato DM. Teaching play skills to young children with autism. J Intellect Dev Disabil. 2013;38(1):74–90.
51. Booth PB, Jernberg AM. Theraplay: helping parents and children build better relationships through attachment-based play. Wiley; 2009.
52. Siu AF. Effectiveness of Group Theraplay® on enhancing social skills among children with developmental disabilities. Int J Play Ther. 2014;23(4):187.
53. Geretsegger M, Elefant C, Mössler KA, Gold C. Music therapy for people with autism spectrum disorder. Cochrane Database Syst Rev. 2014;6.
54. Hillier A, Greher G, Poto N, Dougherty M. Positive outcomes following participation in a music intervention for adolescents and young adults on the autism spectrum. Psychol Music. 2012;40(2):201–15.
55. LaGasse AB. Effects of a music therapy group intervention on enhancing social skills in children with autism. J Music Ther. 2014;51(3):250–75.
56. Vaiouli P, Grimmet K, Ruich LJ. "Bill is now singing": joint engagement and the emergence of social communication of three young children with autism. Autism. 2015;19(1):73–83.
57. Lim HA. Effect of "developmental speech and language training through music" on speech production in children with autism spectrum disorders. J Music Ther. 2010;47(1):2–26.
58. Edgerton CL. The effect of improvisational music therapy on the communicative behaviors of autistic children. J Music Ther. 1994;31(1):31–62.
59. Wan CY, Bazen L, Baars R, Libenson A, Zipse L, Zuk J, et al. Auditory-motor mapping training as an intervention to facilitate speech output in non-verbal children with autism: a proof of concept study. PLoS One. 2011;6(9):e25505.
60. Fancourt D, Garnett C, Spiro N, West R, Müllensiefen D. How do artistic creative activities regulate our emotions? Validation of the Emotion Regulation Strategies for Artistic Creative Activities Scale (ERS-ACA). PLoS One. 2019;14(2):e0211362.
61. Wimpory D, Chadwick P. Brief report: musical interaction therapy for children with autism: an evaluative case study with two-year follow-up. J Autism Dev Disord. 1995;25(5).
62. Kern P, Wolery M, Aldridge D. Use of songs to promote independence in morning greeting routines for young children with autism. J Autism Dev Disord. 2007;37:1264–71.
63. Oono IP, Honey EJ, McConachie H. Parent-mediated early intervention for young children with autism spectrum disorders (ASD). Evid Based Child Health Cochr Rev J. 2013;8(6):2380–479.
64. Schreibman L, Dawson G, Stahmer AC, Landa R, Rogers SJ, McGee GG, et al. Naturalistic developmental behavioral interventions: empirically validated treatments for autism spectrum disorder. J Autism Dev Disord. 2015;45:2411–28.

65. Kasari C, Gulsrud A, Paparella T, Hellemann G, Berry K. Randomized comparative efficacy study of parent-mediated interventions for toddlers with autism. J Consult Clin Psychol. 2015;83(3):554.
66. Liao S-T, Hwang Y-S, Chen Y-J, Lee P, Chen S-J, Lin L-Y. Home-based DIR/Floortime™ intervention program for preschool children with autism spectrum disorders: preliminary findings. Phys Occup Ther Pediatrics. 2014;34(4):356–67.
67. Hardan AY, Gengoux GW, Berquist KL, Libove RA, Ardel CM, Phillips J, et al. A randomized controlled trial of Pivotal Response Treatment Group for parents of children with autism. J Child Psychol Psychiatry. 2015;56(8):884–92.
68. Zhou B, Xu Q, Li H, Zhang Y, Wang Y, Rogers SJ, et al. Effects of parent-implemented Early Start Denver Model intervention on Chinese Toddlers with autism spectrum disorder: a non-randomized controlled trial. Autism Res. 2018;11(4):654–66.
69. Srinivasan SM, Pescatello LS, Bhat AN. Current perspectives on physical activity and exercise recommendations for children and adolescents with autism spectrum disorders. Phys Ther. 2014;94(6):875–89.
70. Escalona A, Field T, Singer-Strunck R, Cullen C, Hartshorn K. Brief report: Improvements in the behavior of children with autism following massage therapy. J Autism Dev Disord. 2001;31:513–6.
71. Getchell N, Miccinello D, Blom M, Morris L, Szaroleta M. Comparing energy expenditure in adolescents with and without autism while playing Nintendo® Wii™ games. Games Health Res Dev Clin Appl. 2012;1(1):58–61.
72. Hildebrandt MK, Koch SC, Fuchs T. "We dance and find each other": effects of dance/movement therapy on negative symptoms in autism spectrum disorder. Behav Sci. 2016;6(4):24.
73. Scharoun SM, Reinders NJ, Bryden PJ, Fletcher PC. Dance/movement therapy as an intervention for children with autism spectrum disorders. Am J Dance Ther. 2014;36:209–28.
74. Koenig KP, Buckley-Reen A, Garg S. Efficacy of the Get Ready to Learn yoga program among children with autism spectrum disorders: a pretest–posttest control group design. Am J Occup Ther. 2012;66(5):538–46.
75. Pan C-Y. Effects of water exercise swimming program on aquatic skills and social behaviors in children with autism spectrum disorders. Autism. 2010;14(1):9–28.
76. Altman N, Briggs R, Frankel J, Gensler D, Pantone P. Relational child psychotherapy. Other Press, LLC; 2010.
77. Alvarez A. Addressing the element of deficit in children with autism: psychotherapy which is both psychoanalytically and developmentally informed. Clin Child Psychol Psychiatry. 1996;1(4):525–37.
78. Bromfield R. Playing for real: the world of a child therapist. Dutton Adult; 1992.
79. Shapiro T. Psychotherapy for autism. J Infant Child Adolesc Psychother. 2009;8(1):22–31.
80. Tustin F. The protective shell in children and adults. Routledge; 2018.
81. Houzel D. The psychoanalysis of infantile autism. J Child Psychother. 2004;30(2):225–37.
82. Reichow B, Barton EE, Boyd BA, Hume K. Early intensive behavioral intervention (EIBI) for young children with autism spectrum disorders (ASD). Cochrane Database Syst Rev. 2012;10.
83. Volkmar FR, McPartland JC. From Kanner to DSM-5: autism as an evolving diagnostic concept. Annu Rev Clin Psychol. 2014;10(1):193–212.
84. Lord C, Elsabbagh M, Baird G, Veenstra-Vanderweele J. Autism spectrum disorder. Lancet. 2018;392(10146):508–20.

GPSR Compliance

The European Union's (EU) General Product Safety Regulation (GPSR) is a set of rules that requires consumer products to be safe and our obligations to ensure this.

If you have any concerns about our products, you can contact us on ProductSafety@springernature.com

In case Publisher is established outside the EU, the EU authorized representative is:

Springer Nature Customer Service Center GmbH
Europaplatz 3
69115 Heidelberg, Germany

Batch number: 10726135

Printed by Printforce, the Netherlands